Praise for Jessica Speart's *Winged Obsession*

"A true thriller about an undercover U.S. Fish and Wildlife agent hunting for the 'human vacuum cleaner' of the insect world. . . . The world of butterfly smuggling nets an estimated $200 million annually, and is filled with unforgettable villains . . . and the fascinatingly weird. *Winged Obsession* swings open the door."

— *Cleveland Plain Dealer*

"Reads like a suspense thriller." — *Oklahoma City Oklahoman*

"*Winged Obsession* offers a fascinating glimpse into the illegal market in exotic and endangered butterflies. . . . I was rooting for creatures I didn't know existed before I read the book." — *Miami Herald*

"Butterfly smugglers? Who knew? Journalist Jessica Speart chases down the butterfly world's most elusive criminal, the notorious Yoshi Kojima, in her fantastic new book. It's a journey with the twists and turns of a taut thriller—like *The Orchid Thief*, only with wings."

— *DailyCandy*

"Meet the Hannibal Lecter of the conservation world. . . . This exposé reads like a thriller and proves once again that truth is stranger than fiction."

— Lee Child, *New York Times* bestselling author

"*Winged Obsession* is an unputdownable thriller that exposes a new world of greed and obsession—the illegal trafficking of exotic butterflies. They're among nature's most gorgeous and remarkable creatures, but they're worth millions of dollars on the black market, and Jessica Speart makes you root for them and Special Agent Ed Newcomer as he wages a high-stakes game of cat-and-mouse with notorious butterfly smuggler Yoshi Kojima. I loved this book!"

— Lisa Scottoline, *New York Times* bestselling author

Winged Obsession

BY JESSICA SPEART

FICTION

Unsafe Harbor

Restless Waters

Blue Twilight

Coastal Disturbance

A Killing Season

Black Delta Night

Border Prey

Bird Brained

Tortoise Soup

Gator Aide

NONFICTION

Winged Obsession

WINGED OBSESSION

The Pursuit of the World's Most Notorious Butterfly Smuggler

JESSICA SPEART

wm
WILLIAM MORROW
An Imprint of HarperCollins*Publishers*

This is a work of nonfiction. The events and experiences detailed herein are based on interviews with witnesses, investigative reports, audio recordings of undercover meetings, numerous hours of undercover video of Skype conversations, and court documents. Some names, identities, and circumstances have been changed in order to protect the privacy and/or anonymity of the various individuals involved.

A hardcover edition of this book was published in 2011 by William Morrow, an imprint of HarperCollins Publishers.

First William Morrow trade paperback edition published 2012.

Designed by Jamie Lynn Kerner

Library of Congress Cataloging-in-Publication Data has been applied for.

ISBN 978-0-06-177244-3(pbk.)

12 13 14 15 16 OV/RRD 10 9 8 7 6 5 4 3 2 1

For those who dedicate their lives to protecting the wild things

Man has been endowed with reason, with the power to create, so that he can add to what he's been given. But up to now he hasn't been a creator, only a destroyer. Forests keep disappearing, rivers dry up, wild life's become extinct, the climate's ruined and the land grows poorer and uglier every day.

—ANTON CHEKHOV, *UNCLE VANYA*, 1897

The least movement is of importance to all nature. The entire ocean is affected by a pebble.

—BLAISE PASCAL

Winged Obsession

Prologue

obsess *vt*: to haunt, fill the mind—obsession *n*

"Hey, what does your guy look like again?"

Special Agent Ed Newcomer stood planted behind a column in the Customs area of Los Angeles International Airport. It was the busiest time of the day. The place looked like the middle of Times Square during rush hour. It was awash with new arrivals in a moving mosaic of sights, sounds, and smells.

He stared in disbelief as Immigrations and Customs Enforcement (ICE) agent Jamie Holt asked the question. The agent couldn't be serious. Newcomer was quickly getting a bad feeling about all of this. "What do you mean?" he asked in return, his stomach doing a three-sixty.

"Well, there are only about five people left, and no one in Immigration has seen him yet," Holt admitted.

Though neither of them said a word, it wasn't a far stretch to guess that there'd been a screwup somewhere along the way.

Newcomer's eyes darted around nervously as he broke into a cold sweat. *How could he simply disappear? Who in the hell was this guy, anyway? Houdini?* "He was on the flight, right?" he double-checked.

This would be the third time that his perp had slipped away. *Three strikes and you're out,* he thought cynically, a wave of nausea beginning to overtake him.

"Yeah, yeah. He was definitely on the flight. Just hold on. Don't worry yet. I'll be right back," Holt said, trying to reassure him before rushing back toward Immigration.

Don't worry. That was a good one. All this case had been so far was one major headache. After three long years of grueling undercover work, this was supposed to be his payday.

Newcomer took a deep breath, trying to summon his last bit of energy to put this case to rest. Panic was already oozing out of him. LAX was the fifth-busiest airport in the world, with fifty-five million people passing through every year. It felt as if each and every one of them was here today.

Holt ran back, looking as if she'd just seen a ghost. "He was here all right, but he somehow got through."

The words shot through Newcomer's brain as if they'd been fired from a .357 Magnum. "What the hell do you mean he somehow got through?" he demanded.

Four hundred people had just disembarked from a Japan Airlines flight. The room started to spin as Newcomer feverishly searched through a sea of Asian faces. The crowd became a nonstop blur of moving body parts and jostling luggage intent on only one thing: making their way as quickly as possible toward the exit.

A bitter laugh began to rise in Newcomer's throat. *Goddamn it. Had Kojima really won again? It was as if he'd taken a play straight out of Bobby Fischer's handbook and just declared, "Checkmate."*

How in the hell had his case ever come to this?

THE L.A. INSECT FAIR

Without obsession, life is nothing.
— JOHN WATERS

IT WAS A PERFECT MORNING. There couldn't have been a better day for the annual Bug Fair at the Natural History Museum of Los Angeles County. It was only 10 AM that May 30, 2003, but a crowd was already spilling down the museum steps and forming a long line on the walkway, waiting to get into the exhibit. The swarm of people resembled a giant wriggling millipede.

"Bugs are cool," a mom told her son as they stood on line for the weekend event.

Who's she kidding? U.S. Fish and Wildlife special agent Ed Newcomer thought to himself. *Has she ever taken a good look at a water bug? They're creepy, crawling, disgusting insects.*

"We're the real pests here," a man informed his daughter reflectively.

A six-foot-tall Timmy the Terminator bug playfully swatted the girl and then handed her an insect coloring book.

If Newcomer had been the father, he'd have taken the coloring book and used it to smack the mutant bug back.

Even celebrities patiently waited in line with their youngsters in tow.

Newcomer slipped past the throng and through the front door to blend with the growing crowd inside. Twelve thousand people turned out for the event, making the weekend the museum's two busiest days of the year. Who would have thought an exhibit of bugs and butterflies would draw so much attention?

The museum's entire first floor was dedicated to the fair. Seventy vendors lined the African Mammal Hall and the North American Mammal wing under the watchful eyes of the stuffed resident wildlife. Every nook and cranny was crammed with insects either dead or alive and crawling. Kids squealed in a mixture of horror and delight as exhibitors placed live tarantulas in their palms and let them crawl up their arms. A bug chef beckoned to passersby to stop and sample his wares. The menu included tarantula tempura, desert hairy scorpion scallopini, and Washington waxworms delicately seasoned with sugar and shredded coconut. It was a rare treat for those brave enough to give it a try. Newcomer wasn't among them.

He worked his way past booths touting beetles elegant as couture jewelry, their shells the color of precious gems. But the main event was the butterflies. There were morphos, blue as Paul Newman's eyes, and birdwings, green as showers of shamrocks. Still others flaunted wings in rainbow hues of purple, sun-drenched orange, blood red, and blacks deep and dark as infinite black holes. Each was pinned and mounted like a fine work of art. Except these masterpieces had once been alive.

The museum seemed the perfect venue for such a show, since butterflies are at least forty million years old. What other species lives almost everywhere on earth but for the frigid Antarctic and the most arid of deserts? Along with their vast territory is their range in size. The smallest species is the half-inch pygmy blue butterfly, while the Queen Alexandra, with its eleven-inch wingspan, is larger than many birds.

However, size isn't the only difference among butterflies. Their life span also varies. The spring azure appears with the first warm weather of spring and lives only a few days, whereas the mourning cloak, with its funereal dark cape of wings fluted in gold, is the longest-lived butterfly, lasting eleven months. It's no wonder there are approximately 20,000 species of butterflies worldwide, of which 725 of them reside in North America. Even more mind-boggling is that each has its very own unique wing pattern.

Though the fair was promoted as an educational event, it was also a big moneymaker for vendors, with wads of cash changing hands. Sure, there were butterflies being sold for five and six bucks apiece. But there were others with price tags of up to fifteen thousand dollars. Newcomer was always amazed at what people would spend on dead wildlife. It seemed they'd become so detached from the natural world that they preferred their nature in a box.

He took it all in as he studied one booth after the next. He wasn't there for the bugs, and though he got a kick out of seeing movie stars, they weren't his prey du jour, either. He glanced down at the photo in his hand. It was a legal resident-alien driver's license, gratis the California Department of Motor Vehicles. His prey was an Asian man who was a notorious bug collector. It was time to make the donuts and find his quarry.

A moment later he spotted his target. Or, at least, that's what he hoped. The man didn't look much like his driver's license photo. But then again, who did? Newcomer's own driver's license made him look like a kid. Still, there could be little doubt. This had to be the guy he was after.

Hisayoshi Kojima stood surrounded by a group of admirers, each tightly gripping his own beautifully polished wooden box—the sign of serious insect collectors. He watched as the men jockeyed for position, hoping to catch Kojima's eye and snag much-sought-after treasures. Kojima smiled with all the esteem of a rock star enjoying the limelight as a steady stream of U.S. currency flew into his fanny pack. The insect dealer was taking in more money than any other vendor. This guy was definitely the Bruce Springsteen of the Bug Fair.

Newcomer had heard that Kojima saw himself as the Indiana Jones of insects. That was all right. Newcomer liked to imagine himself as

Brad Pitt. A man could dream, couldn't he? However, the stocky Japanese national was fifty-three years old, five feet seven inches tall, and had a pudgy face defined by bushy black eyebrows. His thinning dark hair looked as though it had never seen a comb. His complexion was pale as vanilla pudding, and his skin looked as soft as that of the Pillsbury Dough Boy.

He found it hard to believe that Kojima was a daring adventurer, judging by the look of him. That should give any little boy hope. But it wasn't the only way in which Kojima was routinely described. The man with the fanny pack, loose khaki pants, and stained polo shirt was widely regarded in collecting circles as the world's most wanted butterfly smuggler.

This wasn't how Newcomer had planned to kick-start his new career. For God's sakes, he'd become an agent in order to tangle with real bad guys, not some nerd who liked to play with bugs. Newcomer had dreamt of beginning with something much sexier. Going after elephant-ivory dealers or tracking down gorilla poachers would have fit the bill. Give him a case that involved smuggled polar bear hides, an illegal shipment of parrots, or illicit tiger bones to solve, and he would have been happy. Was that too much to ask for? How else was he going to make an impact and draw attention? The powers that be at Fish and Wildlife liked big splashy cases.

Instead, his boss had handed him a freaking butterfly dealer. It wasn't that insects and butterflies were at the bottom of his list. They'd never made it that far. No way were bugs even on his radar.

Newcomer was well aware that he was low man on the totem pole. No one else in the office had wanted the case, and with good reason. The learning curve alone was a killer. Butterflies are all identified by their Latin names. Tackle a case like this, and you were expected to learn them. He'd never been that great a student to begin with, but when you're a rookie, you take whatever you get.

Newcomer's initial hunch about Kojima proved to be correct as another person now sidled over to the booth. It was the man who had called Fish and Wildlife a few days ago. He'd complained that Kojima was illegally catching endangered butterflies and beetles from around the world and smuggling them into the United States. Word had it that he'd just returned from a collecting trip in Bolivia and Costa Rica. He planned to offer his newly acquired booty at the fair. Kojima was selling stuff that no reputable dealer could get his hands on.

Newcomer's assigned task was to babysit the whistleblower, who just happened to be another insect dealer. The guy had agreed to help by turning confidential informant, or CI. It was common practice in the cutthroat and low-down dirty world of the wildlife trade. Clearly, the CI had his own agenda. He probably hoped to push Kojima out of business and level the playing field for himself.

It all came down to tit for tat. Fish and Wildlife had recently penalized this same CI for failing to maintain the proper import permits. The guy had become so pissed that he'd decided to turn the heat on someone else. That was how Fish and Wildlife usually got its best tips.

As far as he knew, most insect dealers tried to follow the rules by filing the proper paperwork and coughing up the import fees. But that wasn't the case with Kojima. Not only did he not play by the government's rules, he proceeded to undercut dealers left and right on even the most common, legal butterflies. The fact that Yoshi constantly got away with it made dealers furious.

Kojima was known to be a one-man demolition derby when it came to collecting. The guy was a genuine environmental nightmare. He managed to acquire endangered butterflies that not even museums or university collections could obtain. His phenomenal merchandise and prices made him the darling at all the insect fairs. Kojima sold out every time. Naturally, he was as popular with other dealers as a hooker at a DAR tea party.

No one could understand why the Service hadn't yet nailed him.

Rumor had it that once Fish and Wildlife had you in its sights, it would take you down, destroy your career, and virtually ruin your life. Tales abounded of jackbooted agents thuggishly kicking in doors and throwing dealers on the floor as their wives and children watched in horror. The nightmare stories grew like an out-of-control game of telephone.

So why was Kojima allowed to run amok? other insect dealers wondered.

It was time to take the guy down.

Newcomer had bitten the bullet and done his homework a few days prior to the fair. What do you know? Kojima had his own Web site, like any other good smuggler. Splashed across its pages were colorful displays of beetles and butterflies just as if they were a wide variety of designer shoes. Even better, PayPal and credit cards were accepted.

Unbelievable. This guy really has it together, Newcomer realized.

Kojima was certainly no dummy. He wisely didn't list prices. But the real showstopper was the mascot that fluttered just below the Web site's banner. This thing had to be the Angelina Jolie of the butterfly world. It was big, beautiful, and splashy. Newcomer didn't care about bugs, and even he found himself attracted to it. The wings were an iridescent bluish green dabbed with flecks of gold, as enchanting as Joseph's amazing Technicolor dreamcoat, and the butterfly was as big as a bird. Newcomer knew it had to be expensive. What was this butterfly anyway, and why didn't Kojima have the species name listed?

Newcomer Googled the butterfly's description, and voilà! He had his answer.

Son of a bitch! No wonder Kojima hadn't bothered to mark down the name. Any amateur collector would have known what it was. Even *he* had heard of it. Queen Alexandra's Birdwing was the largest and one of the most endangered butterflies in the world. It was like slapping the panda icon on a conservation site, except Kojima was probably selling the damn thing! The butterfly had to be worth a fortune.

Newcomer watched as the CI approached and initiated a conversation with Kojima. It was clear that the two men knew each other. The CI wore a hidden recorder and had been well versed on what questions to ask. If all went well, Kojima would eventually take the informant into his confidence.

Everything was going according to plan. But the day was long, the CI, also working as a vendor, had returned to his own booth, and Newcomer was becoming bored. How many frigging bugs could one look at without beginning to feel itchy and have the urge to scratch? He decided to stroll over to Kojima's stall and see for himself what all the fuss was about.

One look was all it took. He had to hand it to the guy. Kojima's butterflies were the finest at the fair. His merchandise looked absolutely flawless. Many of his specimens were so gorgeous they took one's breath away. Newcomer listened to the conversation and then decided he might as well ask a few questions of his own. What the heck? Playing the wide-eyed novice should be a breeze. As luck would have it, he didn't know squat about butterflies. Anything he learned was bound to be a bonus. Besides, he was curious about the man and hoped to get some insight into him.

"What kind of butterfly is that?" he asked during a lull, pointing to a birdwing that was as colorful as a Chagall painting. The wings were neon green, and the abdomen appeared to be freshly dipped in bright yellow pollen.

Kojima glanced at the trim, sandy-haired young man standing beside him. Dressed casually in jeans, a work shirt, and baseball cap, Newcomer looked like your typical Southern California beach boy. A light mustache brushed his upper lip, giving him the air of a kid trying to look older. Everything about him was neat and orderly, but for a cowlick that refused to behave. There was something endearing, yet comical, about it, and Yoshi Kojima found himself intrigued. All in all, this was a nice-looking boy.

"That's *Ornithoptera goliath*. It's very beautiful. You like butterflies?" Kojima inquired in a soft, heavily accented voice.

Newcomer flashed an easygoing smile that made him look even younger than his thirty-seven years.

"I like them. I just don't know much about them," he admitted, purposely conveying an "aw-shucks" attitude.

Kojima liked people with an open heart. This boy seemed to be one of them. "I'm Yoshi."

"Ted," Newcomer responded on the fly. *Whew!* Thank God he'd picked an undercover name during training. That had just saved his butt. "You have some cool-looking butterflies. I'd love to buy a few, but I don't have much money for this sort of thing right now."

"That's all right. You don't need to spend a lot," Kojima attested. Talking about butterflies was one of his favorite things, and he never tired of it. "You have a collection?"

"Not yet. I'd like to do it as a hobby, and you seem to have the best stuff here," Newcomer told him.

"Yes. Mine are drop-dead gorgeous," Kojima agreed without a trace of modesty. He could have gone on for hours about them, but someone else grabbed his attention and he caught the whiff of money.

Newcomer took his cue. He wandered around for a while but was soon drawn back to Kojima's booth. Newcomer was one in a long line of suitors. Collectors were hanging around Kojima like bees drawn to honey. Newcomer patiently waited and then jumped in when he had the chance.

"Wow! That's a neat bug," he said, pointing to a golden scarab beetle.

"You know anything about them?" Kojima asked bemusedly.

It was funny. He'd expected Kojima to be guarded, but the guy was easy to talk to. *That's always a good thing in a smuggler,* Newcomer thought ironically, and laughed to himself. "Nope. Absolutely nothing," Newcomer acknowledged with a grin. "But I'd like to learn, and you clearly know more than any other dealer here."

The boy was smart. "What's your name again?" Kojima asked, his curiosity having been piqued.

Ooh, yeah. Remember that Kojima responds to flattery. Newcomer made a mental note of it. "Ted Nelson," he said.

"Okay, Ted. Here's your first lesson." Kojima patiently proceeded to point out different beetles and butterflies, telling him a bit about each and where they were from, along with their price. One never knew. The boy might turn out to be a good customer.

"You see these?" Kojima fanned his hand over a group of beetles. "They look like jewels, right?"

Newcomer squinted and did his best to ignore their spindly little legs. When they were seen in the right light, Kojima's description couldn't have been more apt. There were bugs in every color of an artist's palette. Some of them looked nearly psychedelic. They could have been conjured by Timothy Leary during one of his acid trips. Still others were as hypnotically metallic as shiny new cars. Their shells reflected the light in tones of deep bronze, plated gold, and highly buffed chrome, as if they'd been dipped in multiple layers of lacquer. Then there were those that resembled gaudy pieces of costume jewelry. If you forgot they were bugs, they were really pretty cool.

"I collect them in Central and South America," Kojima disclosed.

"What about this one?" Ed asked, and pointed to a dead large horned beetle.

The thing could have been the star of its own science-fiction movie. It was the color of rich, deep cocoa, with a head the shade of bittersweet chocolate. But the main attraction were its two long, sharp horns, each shaped like a sleek lobster claw. A closer look revealed thick hairs lining the upper horn that were stiff as bristles on a brush.

"Oh, that's *Dynastes hercules*. It's the largest of the rhinoceros beetles," Kojima replied.

No kidding. The thing was humongous.

Kojima explained that some rhinoceros males reach 7.15 inches in length, with horns that can grow even longer than their bodies. They use them to fight for females and battle other males over territory. The Hercules beetle is the strongest creature on earth for its size, Kojima explained, and can carry 850 times its own body weight.

Newcomer thought about that for a moment. Was he kidding? That had to be the equivalent of a 180-pound man lifting a 60-ton M1 Abrams battle tank over his head. Newcomer looked at the bug again, with newfound respect. Rhino beetles were basically the super-pumped-up, steroid-enhanced Arnold Schwarzenegger version of insects.

"It's from South America, too. I sell live ones in Japan for ten thousand dollars apiece," Kojima bragged.

Newcomer whistled under his breath. Kojima was one sharp dude. "That's a lot of money for a bug. What's that thing like when it's alive?" he asked with genuine interest.

Kojima's eyes nearly twinkled with excitement. He waited until no one was around. Then he reached back and grabbed a small plastic cage covered with newspapers. He motioned for Newcomer to join him.

What's he up to now? Newcomer wondered.

He had his answer as Kojima removed an enormous horned beetle from inside the pen.

"Wow! That thing's unbelievable!" Newcomer blurted in surprise.

It was the same size as the other beetle but with a shorter horn. Oh, yeah, one other thing: It was alive and squirming.

Kojima nodded in delight. "This type of *Dynastes* can only be found in Bolivia. I collected thirty of these just a few weeks ago. I already sold them all in Japan for ten thousand dollars each. Only this one I brought here to the U.S."

Newcomer tried to imagine Kojima running around in the jungle wearing a pith helmet and carrying a net. He couldn't.

Kojima moved forward to place the beetle in Newcomer's hand; Newcomer instinctively flinched. Was Kojima out of his mind? No way in hell was he going to touch that thing.

"Is that legal?" he quickly dodged, hoping to cover his reaction.

Newcomer already knew the answer. Bringing live insects into the country without a permit was totally verboten.

Kojima shrugged, as if the law was a minor inconvenience. "It's illegal. But ninety-nine percent is safe. Sometimes we pay under the table."

There it was—the tiniest smirk. Newcomer picked up on it like a trained bloodhound. Kojima loved getting away with it. This guy wore what he did like a badge of honor.

Someone approached and Kojima swiftly put the beetle away.

"You come back at the end of the day. Maybe I have some extra butterflies I can give you," Kojima kindly suggested, then turned toward his customer.

That sounded like a good deal. Besides, Newcomer had nothing to lose. He slowly made his way over to the informant's booth.

The CI finished up some business and then slipped him a white plastic grocery bag. Inside were the recorder, a microphone, and a minidisc. "I got what I could on tape. What do I do next?" he asked, glancing around nervously.

Newcomer watched in amusement. Didn't these guys ever go to the movies and know how to act inconspicuous?

"Just let me know if Kojima contacts you, and I'll be in touch," Newcomer instructed.

He took another leisurely stroll around the museum floor, stopping at a few more booths. Try as he might, it was hard not to admire the exhibits, regardless of the fact that everything was dead.

Some butterflies were bedecked with gaudy silver spangles as if in a nod to Beyoncé and Cher, while others had full regalia of blue fireworks on their wings. Then there were those that resembled winged angels in long flowing ball gowns. If alive, they might have been miniature pieces of Tiffany glass in swirling tapestries of color. But these butterflies lay perfectly still, tiny corpses in a glass-covered morgue, each with its very own toe tag. The label provided detailed information regarding the species, along with when and where it had been collected—the necessary data that every serious butterfly collector demanded.

Newcomer was engrossed in studying the merchandise when he felt a sudden tap on his shoulder.

Whirling around, he was startled to find Kojima standing behind him. He froze for a split second. *Holy crap! It's him. Kojima must have somehow found out who I am,* he thought, panicking. Terrific. Now the case would be *bing, bam, boom,* over. That ought to make a hell of an impression on his new boss.

Except that Kojima stood holding a plain cardboard box in his hands.

"Here are some butterflies for you," he said with a smile, and offered the gift as if it were a box of Godiva chocolates.

Newcomer was momentarily stunned by the gesture.

"Wow, this is incredibly nice of you, Yoshi," he said. He took the box and opened the lid. Inside were thirteen dead butterflies, each one a perfectly mounted gem.

"It's nothing," Kojima told him. "Just common butterflies. They're not valuable but are good to start your collection."

Their worth wasn't what mattered. It was the fact they came from Kojima that made them as precious as gold. Newcomer would never have dreamt he'd make this kind of connection with the man. "Are you kidding? This is terrific. Please, let me pay you for them," Newcomer said.

"No, no. These are for you. They're free," Kojima said.

"Th . . . thanks, Yoshi," Newcomer stammered, still feeling somewhat dumbfounded. *Snap out of it,* he ordered himself. *For chrissake, take advantage of the moment!* "Is there a way I can get in touch with you?"

Kojima took a second to size him up. No alarm bells went off, and he prided himself on being a good judge of character. Ted Nelson looked genuinely eager to learn, and he was always willing to help those who were truly interested in bugs. Kojima pulled out a black Magic Marker and wrote his e-mail address on the box.

"Thanks again. This is really terrific of you," Newcomer said, gripping the box of butterflies. Wanting to leave before Kojima could change his mind, he made a mad dash for his car.

Catching his breath, he was on a high, unable to believe his luck. It turned out this little interlude with bugs hadn't been so bad after all. He could hardly wait for his next assignment.

He barely bothered to notice the endless line of traffic snaking by as he turned onto Interstate 110. To his mind's eye, each vehicle was as bright and shiny as a tropical butterfly. The traffic remained bumper-to-bumper all the way to his office.

Undercover Man

Passion is a positive obsession. Obsession is a negative passion.
—Paul Carvel

Torrance is a huge sprawling suburb just south of Los Angeles, home to generic shopping malls and an Exxon Mobil oil refinery, and close to the airport.

Newcomer pulled into Torrance Tech Park, a nondescript strip of brick office buildings, and walked toward the unobtrusive sign that announced OFFICE OF LAW ENFORCEMENT. He was "home, sweet home."

He unlocked the front door, entered the lobby with its bulletproof window, punched in the code at the security panel, and stepped inside the honeycomb maze of the U.S. Fish and Wildlife Service office.

This case is going to be a cinch, he thought as he photographed the butterflies, tagged the box, and placed it into evidence. All the CI had to do was follow through, with a little guidance from Newcomer. His mission was over. Maybe now he'd be given something more substantial to tackle.

It didn't matter that it was late Saturday afternoon. He was anxious to report in. He picked up the phone and called his boss.

"Hey, you're not gonna believe what just happened at the fair," he began when Marie Palladini answered her phone.

The resident agent in charge listened to his report. She'd thought of the assignment as simply a training exercise for her latest rookie. He'd only been here a few weeks, and she liked to start new agents off easy. She remembered all too well what it was like to be a rank beginner.

Palladini had been one of the first female special agents to join the Fish and Wildlife Service (FWS) in the seventies. She'd never heard of the agency before. Or, if she had, she'd thought they were just a bunch of hunters and fishermen. It had taken an interview with the Drug Enforcement Agency to learn about it. Apparently, DEA didn't have an opening for her, but Fish and Wildlife was looking for a few good women. Palladini had always been an environmentalist, and the thought of becoming a wildlife investigator appealed to her. What could be better than helping animals while running around and catching bad guys?

She'd joined the Service and soon found herself playing more roles than the average Hollywood starlet. She worked undercover as the Cajun girlfriend of a gator poacher and learned to fend off more than just four-legged toothy critters. Other roles included administrative assistant to an agent posing as a buyer of walrus ivory, with the Alaskan version of the Hells Angels as their target. She also portrayed a Native American and delved into the illegal sale of eagle feathers.

This was all in her first few years on the job. You name it and she did it. Palladini knew what it took to get a case done and could bring the strongest men to their knees. In turn, she was called a wolverine and accused of having brass ovaries. She considered those to be her gold stars.

Palladini hadn't really expected anything to come of today, even though Kojima was a known violator who had been on their radar for years. Even so, cases sometimes take on a life of their own. She immediately came to a decision. Newcomer wasn't the typical new guy. He already had plenty of real-life experience that could easily be applied to the job. She felt confident that he had enough ambition to take the case and run with it. Besides, she had her own reasons for wanting to nail Kojima, and once she got her teeth into something, she didn't let go.

"Guess what? The two of you obviously hit it off, so we're dropping the CI. Congratulations. You're the new undercover guy," she informed him.

Oh shit, Newcomer thought. This was totally unbelievable! The only thing worse would have been if she'd wanted him to play with a bunch of tarantulas.

He kept his mouth shut, but he'd always hated bugs. Even as a kid they'd freaked him out. There was something unnerving about the feeling of creepy, crawly legs skittering around on his skin. He shivered just thinking about it. Maybe it sprang from the huge spiders in his Denver basement that leapt on him as a boy. Whatever the reason, fear of insects was in his blood. He didn't care whether or not it was rational.

On the other hand, damn! He held Palladini in high regard and would do whatever she wanted. It had taken him way too many years to become a special agent, and he still had something to prove. He wasn't about to blow it now—bugs or not.

Newcomer had gladly left his former career behind, lock, stock, and barrel. He'd worked as a lawyer for eleven years, first in Washington State and then in his hometown of Denver, and had nearly lost his mind. Much of it had been spent prosecuting health-care fraud. He'd hated every single minute of it.

His passion wasn't for law but rather for animals. Hell, it had started when he was just a kid. He and his two sisters would take stray cats into their home while their mom was at work. It was the only thing that got him down into the basement. Besides housing scary spiders, the cellar proved a useful way to sneak cats inside.

"Our basement was like the demilitarized zone except for hand-to-hand combat with all the spiders," Newcomer joked. "The cellar was where we'd go as kids and do whatever we wanted."

That was until his mother discovered the ploy. She put an end to the rescue mission once the cats reached an even dozen.

Besides animals, Newcomer's childhood obsession was playing cops and robbers. It went way beyond the norm. The kid was a law-enforcement junkie. His idea of a good time was to sit in his mother's 1969 station wagon, place an old battery-powered lantern on the roof, switch it on, and spend hours making siren noises. He once wrote his sister a ticket for making out with her boyfriend while parked in front of their house. It was as if he'd been born with a white hat on.

Young Ed spent the rest of his free time watching Walt Disney wildlife documentaries on TV. As a teenager, he would get into his Jeep and roam the backcountry in search of coyotes, elk herds, and whatever else might be out there.

Naturally, his role models were a trio of kick-ass heroes. Adam-12 got to catch the bad guys, while his G.I. Joe dolls went in search of Bigfoot. Then there was Caine from the TV show *Kung Fu*, who walked across the old American West, lived off the land, and beat the crap out of bullies who tried to pick on him or anyone else that they viewed as weak. It seemed to Newcomer like the perfect life. In fact, it reminded him of the job that he had now.

He went to college having decided to become a Colorado state game warden. It would be the perfect blending of his two passions. He'd be able to protect animals and still have the powers of a cop.

He threw himself into the required wildlife biology courses and quickly slammed against a wall.

"It was awful. I did just terrible. I couldn't get interested in all the minutiae you have to learn about big wildlife. I wanted to study elk in the field, not in some lab. I didn't give a flying fart what kind of mitochondria was in their poop," Newcomer admitted ruefully.

He sank into a deep funk until spotting a job announcement on the school bulletin board for a USFWS special agent. *What in the hell is that?* he wondered. *Some new kind of game warden?*

The job description read exactly like an advertisement for the FBI, only to protect wildlife. It was just what Newcomer wanted. The catch was the necessary requirement: a degree in either law or wildlife biology. He'd already struck out on one. That left him with little choice.

Fantasy is a great thing. Newcomer now daydreamed of becoming a lawyer like Harry Hamlin on *L.A. Law,* dating a babe like Susan Dey and making a million dollars. Or he could skip the big bucks, fancy office, and the Hollywood babe and work with wildlife. His world now had options.

What he didn't realize was that being accepted as a special agent would be nearly as difficult as passing the bar exam. Newcomer applied to the Service only to be passed over twice. Yet he still couldn't stop thinking about it. He knew he could do the job. He just had to convince everyone else.

The third try proved to be the charm. It was also just in the nick of time. He was literally four days away from the cutoff for entry as a special agent—thirty-seven years old. He managed to slip in under the wire.

Newcomer closed his law books, kissed his new bride good-bye, and spent the next eighteen weeks sweating it out at the Federal Law Enforcement Training Center in steamy Glynco, Georgia. Newly accepted entrants to the Secret Service, the ATF, the IRS, and a handful of other federal law-enforcement agencies joined him. If he'd thought playing Adam-12 as a kid was fun, this reality proved to be a blast.

The Criminal Investigator program was pretty much a rehash of what Newcomer had already learned in law school. He breezed through it without even taking notes. It was the second half of training that really revved his engine. Special Agent Basic School covered surveillance, undercover work, and how to interview a subject. Newcomer turned out to be a natural at it.

Then there were the tactics courses. Logging in at five feet six inches and 135 pounds, Newcomer could be taken for a lightweight, the type to get sand kicked in his face at Muscle Beach. That was where he surprised everyone. He took no prisoners. A martial arts expert with a sixth-degree black belt in tae kwon do, he naturally excelled in self-defense classes. His wife had earned a black belt, too. As fate had it, they met at a tae kwon do tournament. It was love at first throw-down.

Newcomer received a badge and a gun upon graduation. Beyond the usual shoot-'em-up toys, a real gun was something Newcomer never thought he would own. Though his favorite game as a kid was cops and robbers, both of his parents were Mennonites. He'd grown up in a nonviolent home.

He was also given a list of available duty stations from which to choose. He wisely let his wife make the decision. Originally a New Yorker, she veered toward the West Coast and picked California.

An attractive brunette, Allison is small and slight in build and nearly as lethal as Newcomer. She's also equally competitive along with being bright. Allison decided to get her Ph.D. in social welfare at UCLA. The plan was that Newcomer would catch the bad guys and she'd do her best to rehabilitate them.

Newcomer couldn't have been happier with her choice of location. California is the gateway to Central America and the hub for trade with the Pacific Rim, making Los Angeles one of the busiest ports in the United States.

LAX was where banded iguanas were smuggled into the country in fake prosthetic legs, and two Asian leopard cats had been stuffed into a backpack. Tortoises were hidden in Chinese toys, and heroin was found sewn inside shipments of koi fish. Think of something crazy, and it was probably already being done. L.A. was ground zero for the kind of work Newcomer wanted to do. There was plenty of action, and he planned to go after the big stuff. Throw a stone and you'd hit a case. It was where an agent could make a name for himself while doing the job of his dreams.

He had made a pact with Allison before starting the job. Weekends would be family time and not spent doing work. But this butterfly case had come up, and he'd already broken the rule. It couldn't be helped. She would just have to forgive him. He now slipped the informant's tape into the recorder and listened to Kojima's fractured English. What he heard sounded like something out of an adventure book.

Kojima bragged about working for National Geographic as a jungle guide. His task was to find and provide bugs for their films. He'd discovered a new subspecies of *Dynastes* beetles during one trip and had caught thirty-four pair of them. Touting his job was how Kojima smuggled pricey insects out of Bolivia.

"Officials at the airport know I work for National Geographic. That's why they almost never check me," Kojima boasted to the CI.

Sometimes he even hid bugs in his jacket. "They asked me once, 'Do you have anything?' The guy's so stupid, he just checked my outside pockets but not the ones inside," Kojima revealed with a laugh.

If he had checked the inside pockets, the inspector would have found a giant Hercules beetle snugly wrapped in a washcloth. Once inspectors did find a live bug inside his luggage. They let both Kojima and the beetle go free. Other Japanese dealers who got caught weren't so lucky. They were immediately deported.

Kojima wove tales of having owned butterfly farms in Indonesia and Costa Rica and said he could get any butterfly from anywhere in the world. Listening to him was better than some of the TV shows that Newcomer used to watch.

The CI asked if Kojima ever worried about Fish and Wildlife catching up with him. Kojima replied that an agent had once come to his L.A. home but that he'd easily been outfoxed.

It was the wrong thing to say. Kojima had just unknowingly issued a challenge. Newcomer made up his mind then and there that he would be the agent to nab him.

He drove to work Monday morning and immediately reported to Palladini, filling her in on the tape.

She listened to Kojima's tales of smuggling and his boast of having duped a federal agent. The story had made the office rounds a few years ago. Palladini knew exactly who it was that Kojima was talking about.

Throughout the 1990s Kojima regularly caught protected butterflies along the northern rim of the Grand Canyon, as well as in other U.S. national parks. *Papilio indra kaibabensis* and *Papilio indra martini*—the names sound as exotic as foreign locales or designer drinks, but they're desirable American swallowtails that are much in demand in Japan.

Kojima was also smuggling a steady stream of butterflies from Mexico into the United States at the time. That put other collectors and dealers under the spotlight as well. They decided it was time to teach him a lesson. Fish and Wildlife special agent John Mendoza was tipped off. But that wasn't the only reason why Kojima was turned in.

Kojima had been hammering away at California butterflies that could be legally caught and sold. The man was like a lethal Energizer Bunny. Once he started to catch them, he just couldn't stop himself. They were little flying dollar signs, and he wasn't about to let any of them go. He caught so many of one species that he temporarily wiped out their population. That proved to be a bad move on his part. It wasn't just the butterfly that was hurt but other collectors, who came away empty-handed. Kojima quickly earned the reputation of being a butterfly hog.

That was the tale with the Apache fritillary, one of California's largest and most spectacular butterflies, which resembles a stained-glass window in flight. The species has a restricted range on the Sierra's eastern slope. Kojima went in, scoured the territory, and caught five hundred of the butterflies in only two days. Then he shipped them all off to Japan, where they sold for thirty bucks apiece. There were no more Apache fritillaries to be found for the next two years. Enough complaints were lodged that the area wound up being permanently closed to everyone. It proved to be another black mark against Kojima. Burn enough collectors, and they're bound to turn on you.

Even so, that wasn't a problem for Kojima. He simply moved on to new territory. He instantly gained notoriety in Payson, Arizona, where the Grant's rhinoceros beetle, the largest bug in the United States, appears.

The beetles live in miles of protected forest surrounding the town and fly into Payson for two weeks in August each year for a raucous convention. The insects are drawn to the gas-station lights at night and literally flood the streets. People gather from all over the world to witness the event and collect as many of the bugs as they can. But nobody could compete with Kojima. He not only claimed the location as his own, he also paid locals to help him. He was so competitive that he fought with people over the beetles and even outran small children as they tried to catch them. No way would he willingly give up a single one. Kojima then mailed the bugs off to Japan to be sold as live pets, which is illegal without federal permission. Naturally, it was something Kojima never bothered to get. People called him a human vacuum cleaner. Others said that he was totally obsessed.

Enter John Mendoza, the quintessential dogged federal agent. All he needed was a trench coat and he would have been the Columbo of protecting critters. He learned of Kojima and set his sights on stopping him.

Mendoza went about the usual routine of setting up surveillances as he tried to catch Kojima collecting on federal land. He even knocked on Kojima's door and interviewed him at home. That was followed up with a letter that clearly spelled out the law. About all that did was to give Kojima a good laugh.

Higher-ups got tired of chasing after a butterfly guy with a net. It was costing them time and money. Mendoza was ordered to shut the case down after eight months without success. That pissed him off, along with the rest of the special agents.

What they didn't know was that Kojima changed his ways after that. He went on to bigger and better things, transitioning from poaching in U.S. national parks to trafficking in endangered insects around the world. So far, Kojima was winning the war and Fish and Wildlife was batting zero. That galled Palladini to no end. Catching him had now become a matter of professional pride. She could have cared less if Kojima dealt in bears, birds, or butterflies. The species themselves made no difference. It still added up to a wildlife violation.

According to Interpol, the illegal worldwide animal trade rakes in a cool ten to fifteen billion dollars a year, and in the world of butterflies, Kojima had established himself as a kingpin. A seasoned smuggler, he'd been breaking the law, catching and selling endangered butterflies, for close to twenty years. He'd even been brazen enough to advertise his merchandise on an insect Web site in the United States. Word had it that the illegal butterfly trade alone was worth two hundred million dollars each year. That meant Kojima had been getting one hell of a piece of the pie.

"Keep up the good work," Palladini told Newcomer as he displayed the butterflies that Kojima had given him.

Terrific, he thought, sure that he'd caught a few of his fellow agents snickering. He was right. Newcomer was soon jokingly referred to as "the butterfly cop."

He spent the next few weeks immersed in everything about bugs and butterflies. He clicked on insect Web sites, spoke to experts, and stayed up nights trying to learn the difference between *Ornithoptera paradisea, Troides helena,* and *Papilio hospiton.* There are about twenty thousand known species of butterflies. That provided twenty thousand different ways for Newcomer to lose his mind. It made cramming for exams back in law school seem easy. But he gradually mastered some of their names and markings, along with an idea of price. Like it or not, he was earning his reputation as "the butterfly cop."

Last, he set up a covert e-mail address for his undercover persona, Ted Nelson. All he needed was a cloak and dagger, and he would have felt like an actual spy. Newcomer was as prepared as he was going to be for now. It was finally time to contact Kojima.

As Ted Nelson, he sent Kojima an e-mail reminding him of where they'd met and of the butterflies that Kojima had given him. He had tried to identify all thirteen and was wondering if he'd gotten them right. Would Kojima mind taking a look at his list?

He remembered that Kojima responded to flattery. Perhaps playing the eager young novice would appeal to the master smuggler. Newcomer purposely misidentified more than half the butterflies, hoping to give Kojima a reason to write back. He cleverly sweetened the pie by mentioning that he'd like to add more butterflies to his new collection. He signed his name Ted Nelson and included his phone number along with a mailing address. The number was an untraceable line at work.

Kojima immediately took the bait. He called later that day believing it to be Nelson's home number. Newcomer had already vamoosed, so Kojima left a message. He'd reviewed the list of butterflies and would gladly go over them. Ted should just give him a call.

Newcomer was over the moon as he listened to his voice mail the next morning. Man, but he was good at this! He gave himself a mental pat on the back. Everything was coming together exactly as planned. This was his very first case, and he intended to knock it out of the park.

He promptly returned the call. Kojima couldn't have been more cordial. He congratulated Nelson on his first identification attempt. There had only been a few mistakes. Exactly what did Ted do for a living, anyway?

Newcomer was fully prepared with an answer. "My dad had a wholesale marine-supply company. I took over the business when he retired. I work mostly at home, but I've got to tell you that I'm bored out of my gourd."

Kojima chuckled. Nelson had a sense of humor. He liked that in a young man.

"I've got a little extra money, and I'm looking for something else to do," Newcomer added. Ted Nelson was hoping to find a good investment. "What about you, Yoshi?"

Kojima revealed that besides butterflies, he also sold Japanese antiques. "I have a very good business. Sotheby's and Christies' are always asking me to sell through them. You wouldn't believe how many Japanese antiques I've found here in the U.S. I think a lot of soldiers must have brought antiques back home with them after World War II."

However, Kojima had done many more things than just that. He was a true entrepreneur. He'd once owned a travel agency, but the business had gone sour after 9/11. It was then that National Geographic had approached him, asking him to be an insect "headhunter." "My work is very much like what Harrison Ford does in his Indiana Jones movies," Kojima joked.

Newcomer laughed in agreement. Yoshi Kojima was a lucky man.

"Maybe you should think about doing something different for a while," Kojima suggested helpfully. A change in career might pay off. It certainly had for him. He not only enjoyed what he did but earned good money at it. He'd made ten thousand dollars during this last L.A. insect fair. Would Ted like to meet for coffee and they could talk about a possible business idea that he had?

Was Kojima kidding? This was fantastic! Newcomer bit like a fish lured with fresh bait. He couldn't believe his luck. Kojima was actually beginning to act like his mentor. Just look at the information he'd managed to gather so far. This case was going to be even easier than he'd originally thought.

Newcomer spent the next few days buckling down to prepare for the meeting. There were still umpteen more Latin names of bugs to be learned. They ran through his head like a continuous loop of film. And he had to make sure that his cover was rock solid. He'd heard horror tales of undercover agents having wallets ripped from their hands and the contents scrutinized for ID. One perp had gone so far as to grab an agent's cell phone and examine it. Newcomer didn't want to take any chances. He made a checklist and kept going over it.

He already had a driver's license in Ted Nelson's name. He now proceeded to build a wallet in his undercover persona. But he felt there was something missing. Then he hit upon it: odds and ends to throw in the car's backseat. A few telling details would add a nice touch of authenticity. All he needed was a little help from a commercial boat dealer.

He stopped by a shop and explained the situation. He planned to use the profession as a cover. Would the owner mind talking to him about it? The proprietor couldn't have been more thrilled if he'd been cast in a movie.

Newcomer left armed with a pile of catalogues on boat parts, along with a teacher's manual for the Marine Mechanics Institute. He slapped some Post-it notes in each brochure and stapled his brand-new business card on the front. Then he tossed them into his undercover vehicle, along with a few greasy rags. He even threw some junk into the glove compartment and trunk.

His meeting with Kojima was set for June 23 at a Starbucks on Venice Boulevard. Newcomer slipped out of his usual freshly pressed pants and crisp, clean shirt and into his new persona. Ted Nelson was a casual kind of guy. That called for a polo top and jeans. A small digital recorder fit neatly into his pocket. He remembered to remove his wedding ring just as he walked out the door. Lesson number four: You never want a target to know you have a family.

He was still new at all this, so Palladini decided to tag along. She waited until Newcomer went inside and met up with Kojima. Then she sauntered in and planted herself on the opposite end of the room.

"You look nice today," Kojima said as they sat down.

He couldn't say the same for Kojima. He looked as rumpled as a heap of old clothes that had been dumped at the laundry. Come to think of it, wasn't he wearing the same shirt he'd had on at the fair? He wasn't exactly a fashion plate. His fanny pack covered the elastic waistband of his shorts, and a baseball cap was slapped haphazardly on his head.

"Tell me more about your business," Kojima said, and leaned toward him.

Newcomer elaborated on the cover that he'd already begun to weave. "As I said, H&N Enterprises belonged to my dad. He was a doctor but always loved boats. That's why he went into the wholesale marine business after he retired. He eventually decided to get out of it and passed the business on to me. It's funny, though. What my dad really loves are butterflies. In fact, he was jealous when he heard of the ones you gave me," Newcomer smoothly recounted, growing comfortable. "Hey, how did National Geographic first hear about you, anyway?"

"Oh, they found me at the insect fair here in L.A.," Kojima replied. With that, he was off and running. "I get paid seven thousand dollars for fifteen days work. Now they want me to move to Washington, D.C., so that I'll always be available for them."

"That's terrific," Newcomer replied, although it struck him as odd. For a smuggler, this guy really liked to talk about himself.

"There's so much pressure, though!" Kojima said. "But they'll pay me fifteen thousand a month to go to Central and South America, and give me an apartment and a car."

Add to the list that Kojima also liked to brag. Even so, Newcomer couldn't help but be impressed. Kojima was obviously damn good at what he did.

Over the course of the next hour, he learned that Kojima excelled at any number of things. Not only had the man owned a travel agency, but he'd also raised Siamese betta fighting fish. He had bred them with fancy half-moon tails and once been awarded the show fish world championship. Afterward, he had immediately lost interest in them and never even bothered to pick up his winning fish from the contest.

"I had a beautiful big house in Mount Olympus up in the Hollywood Hills. A two-million-dollar mansion," Kojima boasted. "One of the bedrooms held two hundred fish tanks. It made the room so moist that the ceiling tiles started to fall. I think I had ten thousand betta fighting fish at one time." He had gotten rid of every single one of them.

For a while, Kojima raised cockatoos and parrots and had thirty birds in his backyard greenhouse. He even taught his African gray to say dirty words in Japanese. But he soon tired of them as well, and moved on to breeding orchids. Kojima would work at things until they reached perfection, only to suddenly grow bored and quit. His passions peaked and ebbed like a roller-coaster ride. But he always succeeded at whatever he did. Friends liked to say that whenever he fell, he'd get back up with a dollar in his hand. They joked that he had Jewish blood in him.

The only thing that continually held his interest was butterflies. Like a doleful lover, he always returned to them.

Kojima's personal life had been tumultuous. He had been estranged from his American wife for years. He quipped that he'd kicked his wife out of the house after she cheated him of money. She'd been pregnant with their child at the time. His son, Ken, was now twenty-three years old, and Kojima was still close to his father-in-law, who was very rich and had a couple of homes in Southern California. Kojima lived in Kyoto with his mother these days but kept a small apartment in L.A.

"It's so easy to breed fish, and beetles, and birds. I collect beetles as a hobby and then sell them to the world market. The Japanese will pay a lot of money for them," he divulged freely.

If Kojima had been dropping bread crumbs, Newcomer would have gobbled them all up by now. He hung on Kojima's every word.

Kojima could teach Ted to mount if he was interested. It was simple. In fact, Kojima would give him some equipment that he'd planned to throw away. He had hundreds of butterflies that needed mounting. Why, he'd even be happy to help Ted learn to breed and sell bugs.

"Do you think you would like to sell butterflies?" he asked.

Yes! Newcomer screamed in his brain. "Absolutely! I hate the boat business. I want to do what you're doing, Yoshi. You know, be an insect dealer," he replied. This meeting couldn't have been going better if it were a dream.

"You can come with me to Costa Rica or Honduras sometime. I'm going next week," Kojima offered, as if it was an everyday occurrence.

Kojima was a real jet-setter. Hell, he probably even traveled on free air miles.

"Oh, yeah? What will you be doing there?" Newcomer asked, aware of the tape recorder chugging away in his pocket.

"I collect silver and gold scarab beetles. Very large ones come from those countries." Kojima couldn't have been happier if he was discussing gems. In fact, these were far better.

"Can you bring them into the U.S.?" Newcomer asked, wanting to get his response on tape.

Kojima gave a noncommittal shrug. "Not through Mexico. If there's a problem, you could go to jail. I take them first to Japan and then bring them here. U.S. Customs is not catching Japanese. We're not so much problem people. They never ask what I bring in. They're always on the lookout for Mexicans," he said with a laugh.

Kojima had every angle figured out. Newcomer was already itching to bring him down.

Kojima next entertained him with tales of close calls. He'd had "big trouble" in Mexico just two months earlier when officials discovered two hundred live beetles in his bag. They questioned his stay of only a day. Kojima explained it away with two magic words—National Geographic.

"They called the top guy there and talked on the phone for two hours. They finally let me go." Kojima flashed a conspiratorial grin.

In addition, he took all two hundred beetles with him.

Another time, a Bolivian official spotted a beetle's sharp horn sticking out of his camera bag. Kojima deftly said it was a butterfly, slipped him a hundred bucks, and skedaddled. Interest in his bag continued when he placed it in the plane's overhead compartment.

"The bugs smelled like pee-pee. Everyone kept looking around wanting to know where the stink was coming from!" Kojima said in amusement, still getting a kick out of that one.

His adventures didn't end there. A beetle once escaped from his bag and began to fly around the plane. It looked like a mechanical wind-up toy with wings that were furiously flapping. Kojima quickly reached up and grabbed it in midair, his fingers working as swiftly and precisely as those of a Shaolin monk. He made sure to tuck those babies in better from then on. And to be much more clever in the future.

Exportation of wild beetles and butterflies is illegal in Costa Rica, and officials are rigorous in their inspections. Kojima solved that problem by bribing a local museum staffer. The staffer still packs Kojima's merchandise and sends it under the auspices of the museum.

"I can help you get bugs on our South American trip. Then we can take them back to Japan together. Don't worry. It won't be a problem. They don't show up on X-ray," Kojima assured him. "Besides, it's not drugs. So who cares?"

Newcomer nearly laughed out loud. Wait until Kojima discovered exactly who he was dealing with.

Kojima became more animated by the minute as he continued to outline his plan. "I can teach you how to sell. Maybe you can be my agent here in the States," he proposed.

Bingo! That suggestion caught Newcomer's ear. "Oh, yeah? Just how would that work, and what kind of money would I make?"

Kojima smiled. He knew that he'd just caught Ted Nelson hook, line, and sinker. He explained that he had a number of Internet auction sites in Japan. Did Ted know how to use eBay?

"Yep. I sure do," Newcomer said.

Kojima's eyes lit up in delight. "Then we can really make money together. My friend is making seven thousand dollars a month! He buys butterflies from me and then sells them on eBay. Sometimes a five-cent butterfly will go for as much as a hundred dollars."

Son of a gun. No wonder the butterfly business was booming. These guys were actually making real money.

Kojima nearly burst with excitement as his scheme unfolded. "I'll bring you lots of things, and you can post them for me on eBay. You can set up an account, and I'll sell my material there. Even with a fifty-fifty split, we'll do very well."

Damn! Was he imagining this, or was Kojima really asking him to be his business partner?

"Why don't you just do it yourself?" Newcomer asked, wondering where all this generosity was coming from.

"I can do so many things, but eBay I cannot figure out. It must be a problem with my English," Kojima replied innocently. "Besides, Fish and Wildlife are looking for me. Only they cannot catch."

Once again, Newcomer was taken aback. Kojima was actually bragging about being a smuggler.

"Why not?" he asked, having pondered that question himself.

"Because there's no evidence," Kojima disclosed. "My L.A. address is not in my name. I also have two passports. One is Japanese and the other is American. That one has a completely different identity."

Newcomer's adrenaline began to thrum. Kojima was brilliant! He could actually write a book about this stuff. "You're kidding! What name do you use?"

Kojima's smile was as sly as the Cheshire cat's. "It's a secret. No one knows my American name yet."

Maybe not, Yoshi. But I'll bet you two to one that I find out, Newcomer thought.

"But won't I get in trouble if I post butterflies for you on eBay?" Newcomer asked, continuing to play the naïve protégé.

"Don't worry. They can't catch you," Kojima assured him. "The account won't be in my name."

Unbelievable! Kojima spun more webs than a spider. It was hard to keep up with this guy.

"I don't get it, Yoshi. If bugs are no big deal, then why are the cops after you?" Newcomer pressed. *Come on, Yoshi. Give it to me blow by blow. Spill your guts for the recorder.*

"Oh, those guys are crazy. They have all these stupid rules and regulations. But nobody really cares," Kojima scoffed.

He then laid out his master plan. "We'll start by selling common butterflies. I have very nice ones. Everything will be perfectly legal at first. That will give us time to figure out who the good customers are and which of them are interested in protected CITES material. Then we can start to sell it to them." He politely folded his hands in his lap as if he had studied etiquette with Emily Post.

CITES, the Convention on International Trade in Endangered Species of Wild Fauna and Flora, is an international agreement that's been ratified by 175 nations. Created in 1973, it aims to regulate the commercial trade of plants and animals that are bought and sold around the world. Consumer demand for these species is so high that things could easily get out of whack and lead to their uncontrolled exploitation. It is at CITES meetings, held every two years, that member nations decide which species have become sufficiently threatened to require protection and which have adequately recovered to be consumed once again by international trade.

Within CITES, two opposing camps wage a voracious battle. The use-it-or-lose-it camp—led by sustainable-use countries such as Japan, Zimbabwe, and Norway—argue that all wildlife has a monetary value and should be used as a resource. The other camp—championed by the United States and environmental groups—worries that CITES is far more interested in trade than in saving species and fights for whatever trade restrictions it can.

The convention relies on a rating system based on what's known as appendices. These determine the level of protection that individual species receive.

Appendix I bans all commercial trade in species such as rhinos, mountain gorillas, giant pandas, Asian elephants, tigers, snow leopards, and blue whales, along with all other plants and wildlife threatened with extinction. That means their illegal sale can bring big bucks. As a result, there are rumors of vaults stockpiled with illegal rhino horns and elephant tusks in anticipation that these species will one day disappear forever.

Appendix II covers those species that are not endangered but considered potentially vulnerable. Trade in them is regulated to assure their continued sustainability. Some species protected under this appendix include the American black bear, Hartmann's mountain zebra, gray parrots, toco toucans, green iguanas, and southern fur seals.

Permits are required for their import and export in and out of most countries. It's a system that depends on international cooperation in order to succeed. However, each nation has its own enforcement laws, along with penalties for violating them, which are not upheld in many cases.

The politics of saving species has become a down-and-dirty game in which animals are the ultimate pawns. CITES is indeed a trade treaty in which enterprises as diverse as the ornamental-fish industry, Ringling Bros. and Barnum & Bailey Circus, and environmental groups all have a stake.

More than thirty thousand species are covered under the treaty, making implementation quite a task. The list includes twelve butterflies, with proposals to protect more currently under way. An additional nineteen butterflies are protected under the Endangered Species Act.

Think of CITES as a worldwide emergency room for wildlife.

However, in reality, the treaty is a bureaucratic nightmare of paperwork, with enough holes to blast the space shuttle through.

Kojima apparently knew every single one and shrewdly took advantage of them.

"Most dealers only care about money and not helping people. But

I'm not that way. I like to help those in the insect business, and you're a nice boy," Kojima said to his new protégé, beguiling him. "Why don't you set up an eBay account, and we'll begin to work together. You can handle the orders, and I'll supply the merchandise."

Had Newcomer been anyone else, he would have been blown away. He eagerly agreed to begin immediately.

Kojima stood up and adjusted his fanny pack. "Come with me. There's something I want to give you."

Newcomer followed him out to the parking lot.

"Which one is yours?" Kojima abruptly asked, and began to scan the vehicles.

Now what's he up to? Newcomer thought. He pointed to a white Chevy Tahoe. "That's mine right over there."

"Oh, what a nice car," Kojima replied. He scuttled toward it. Bending low, he emitted a grunt and scrupulously studied the license plate. Then he began to carefully inspect the grille.

Son of a gun! This bastard isn't as trusting as he pretends to be, Newcomer realized, his pulse beginning to race like a V-8 engine. He knew what Kojima was up to. He was looking for hidden emergency lights. What the hell had set him off?

Small beads of sweat formed on the back of Newcomer's neck, and his fingernails nervously dug into his palms. *Calm down! Maybe he's paranoid enough to do this with everyone.* But Newcomer was having a tough time convincing himself of that as he worked hard to control his breathing. This felt like every damn movie he'd seen where the good guy was about to be discovered.

Play it cool! He's not going to find anything. The Tahoe is clean. But Kojima's mind game was definitely getting to him.

"Such a beautiful car," Kojima murmured, continuing to examine every single inch.

Jesus Christ. He's doing everything but kicking the damn tires, Newcomer silently griped while watching him closely as a hawk.

There was only one reason for Kojima to be doing this. *He's scoping me out to see if I'm an agent. What did I do to tip him off?* Newcomer racked his brain. And what could he do to rectify it now? Then he remembered a lesson he'd learned at special agent basic training back in Glynco: *Ignore unusual behavior, and that's a very good clue that you might be an undercover agent.*

He didn't waste another moment. "Hey, what are you doing, Yoshi?" he asked, having decided to play dumb. "Are you thinking of buying a Tahoe?"

"Oh, I just have to check and make sure you're not a Fish and Wild guy," Kojima replied casually, then turned his attention to the vehicle's interior.

Goddamn it! Kojima was now searching the Tahoe for a police radio. A string of doubts began to eat at him. Could there be anything inside that might give him away? Newcomer hadn't expected such a thorough examination. He had to act fast and take charge of the situation. "What? You've got to be kidding me. You're crazy, man. This is total bullshit," Newcomer said with a shake of his head and began to walk away.

Kojima turned to him with a smile. The boy had passed the test. There was nothing to worry about. Besides, Ted Nelson looked more like a choirboy than a law-enforcement agent. "Everything's okay. Let's go over to my car."

That was too close for comfort, Newcomer thought, tensely plucking at his shirt. Sweat had begun to make it stick to his chest.

He trailed Kojima to a large Toyota pickup with enough dents on it to have been in a stock-car race. Either Kojima had bought it that way or he was one hell of a lousy driver.

Kojima began pulling a number of mounting racks out of the back. "These are a gift for you," he said. Newcomer unlocked his vehicle and Kojima placed them inside. "I'll show you how to mount butterflies next time. I can also give you a few to practice on."

That seemed a good guarantee that there'd be a second meeting.

"One more thing," Kojima said. "This is for your father." He handed

Newcomer a framed and mounted butterfly. "Now he has one, too. It's so he doesn't have to be jealous of you."

What a nice guy, Newcomer thought, momentarily touched by the gesture. Then he took a closer look at the merchandise. It was a common butterfly that Kojima probably wanted to get rid of.

"We'll talk soon," Kojima promised, and drove off.

Newcomer watched as the red pickup disappeared in a sea of traffic.

Yep. I was right. This case is going to be a piece of cake.

Even so, he was still astounded at the way it was unfolding. On the other hand, it was beginning to make sense.

Kojima was on the hunt for someone with little knowledge of butterflies but interest in the trade, who also happened to know how to use eBay. As Ted Nelson, Newcomer perfectly fit the bill. Kojima had stated that there were two clear-cut requirements for his partner in crime. The first was that his eBay associate have a U.S. address. The second was that he didn't have a history in the bug trade. Newcomer's cover neatly met both prerequisites. *What a crafty guy,* Ed realized, and laughed to himself.

There'd be no reason for Ted Nelson to immediately attract Fish and Wildlife's attention. After all, he was brand new at this. At least a year would go by before someone like Ted probably landed on their radar. That gave Kojima plenty of time to gather a new base of customers. He could do a lot of business on eBay until then. In the meantime, Nelson would be cooking his own goose, compliments of Kojima.

"So, how'd it go?" Palladini asked, having appeared by his side.

Newcomer didn't have to think twice. "We're going to get this guy by the end of the summer."

"That's great," Palladini replied, impressed by her new agent's confidence.

"One more thing," Newcomer said. "Kojima's setting me up. He plans to make me his fall guy."

The Collecting Obsession

[Book] collecting is an obsession, a disease, an addiction, a fascination, an absurdity, a fate. It is not a hobby. Those who do it must do it.

—Jeanette Winterson

Obsession is a strange thing. It's a lust that burrows inside, grabs hold, and doesn't let go. A fever gets into your blood that drives you crazy. The desire—whether it's for antique cars, Tiffany lamps, or Barbie dolls—is unquenchable. For the person in its grip there is no escape.

There's always another item to be obtained, whatever the cost. The price can be losing your job, your house, or even your spouse. The compulsion is so great it's like an adrenaline rush. Nothing else matters but the next treasure to be added to your collection and the feeling that it will somehow make your life perfect.

That's the way it is with butterflies. The sight of a silver-bordered fritillary can send a collector into quivers of delight. A rare birdwing is a wondrous thing to behold. To obtain an endangered butterfly becomes one's Holy Grail. For a collector, butterflies can be as dangerous and addictive as any drug.

Ego, drive, and fanatical obsession are the traits needed for the single-minded pursuit of this goal. Butterfly collectors are comparable to hunters lusting after a trophy to nail on the wall. Except in this case the trophy is to be pinned and mounted in a box.

"The passion and excitement of the chase are almost as important as the bug itself," offers butterfly dealer Greg Lewallen. "It's like big-game hunting, only our trophies are under six inches instead of a nine-foot-tall Kodiak bear or a two-thousand-pound bull elephant. Believe me, the weirdest bunch of people you'll ever meet are bug collectors."

Their lives can be reminiscent of a Greek tragedy, with many turning to alcohol and drugs to try to subdue their demon. Collectors have been known to go bankrupt in order to underwrite their butterfly trips. To possess what's rare and forbidden becomes their main mission in life.

"We're driven toward something that doesn't put food on the table and that we can't take with us," concurs one lepidopterist. "It's not all happiness and joy. There's a dark side here."

A composite of an obsessive collector was provided by another lepidopterist, who also preferred to remain anonymous. "He's lost his eighth job in three years due to calling in sick and taking off on collecting trips, and he's depressed because no one else will hire him. As a result, his wife has finally left him. Even so, he still can't control his obsession. Instead, he's become hooked on speed, taken out mortgages, and run through his family's savings while still chasing after butterflies."

"We're a dysfunctional lot," admits University of California at Riverside entomologist Dr. Gordon Pratt. "There hasn't been a day in my life that I haven't had a complete fascination with butterflies. It's truthfully an addiction."

Get personally involved with an avid butterfly collector, and you'll always come in second. Mainly because many of those obsessed with the insects tend to have trouble connecting with people and forming relationships. Butterflies become their refuge. For others, such as writers and artists, butterflies become their muse.

Lewis Carroll, best known for having spent time with Alice, was a butterfly enthusiast. Another famous lepidopterist was *Lolita* author Vladimir Nabokov. Not only did he discover and name dozens of species of butterflies, he kept a collection of male butterfly genitalia coated in glycerin, and methodically labeled, in his office. Little Karner blue butterflies were his major passion. Just an inch in size, the gossamer-winged males are iridescent violet-blue with black rims and delicate white fringe, while the females boast orange crescents along their borders. Nabokov once described the butterflies as covering their habitat in "a sea of blue." They are now an endangered species.

Nowhere is the passion for insects greater than in Asia, and Japan is the mecca. This is the heart of the beetle and butterfly craze. According to an article in *The Economist,* one of every ten Japanese men is considered to be an obsessive butterfly collector. Even the former justice minister belonged to a butterfly club.

Insects are viewed as an integral part of Japanese culture. To the Japanese, beauty and harmony are to be found in natural things. Butterflies and bugs are ingrained in their proverbs and sayings, their legend and lore. Some beetles even symbolize samurai warriors. The Japanese see insects as creatures of beauty rather than creepy-crawly bugs. People keep crickets in their homes because they love to hear their chirping at night. Beetles have become more popular in Japan than even cats and dogs.

Vending machines offer live beetles for sale as pets in the summer. You can buy them just as you would a soda or candy bar. Insect cages and butterfly nets are also available in most stores. Have a sudden hankering to get a bug while you're out shopping? No problem. Southeast Asian rhinoceros beetles can be purchased at Japan's version of Home Depot. It's "beetlemania" to be sure.

Little boys begin their pastime by catching beetles in the forest. The bugs are a source of enjoyment and play. Kids pit them one against the other in daily contests. The bugs battle it out to prove which of them is stronger. This childhood passion for insects only seems to grow. So does the amount of money that's spent on finding rare butterflies and bugs.

A reporter with Japan Broadcasting Corporation travels with a friend each summer on collecting trips to Myanmar. They hire an army of thirty locals to accompany them into the forest, to serve as guides and help carry supplies but also to protect the two men from guerrillas. The friends search for butterflies in territory in which most people are afraid to go. They quickly gather whatever butterflies they can and bring them back to Japan to rear.

Some Japanese collectors take trips not nearly as legal and end up paying the price. A few collectors have spent more time than originally planned on excursions in India and China—getting thrown into prison for illegally collecting butterflies and trying to sneak their contraband out of the country. Those who make it home have more than great adventures to tell. They sell their booty on the Internet for high prices. Some hybrid butterflies go for sixty thousand dollars or more. "That kind of money doesn't even raise an eyebrow," states dealer Greg Lewallen.

Then there are those who prefer to stay home and have others do the footwork for them. Their only requirement is that specimens be flawless. One private collector paid a group of able young catchers to parachute onto a South Pacific island and stealthily gather endangered butterflies for him.

Butterfly collectors with the financial means will do whatever is necessary to obtain the specimens they want. For some, it's the equivalent of collecting a Renoir or Van Gogh. Butterflies and bugs are considered high-end art. For others, it's more like stamp collecting. Dealers know this and will go to extreme lengths to satisfy their customers' appetite. Rumor has it that Kojima found the largest rhinoceros beetle of its kind and sold it to a Japanese collector. The man not only paid one hundred thousand dollars for the bug, he flew to Ecuador and met with Kojima to take possession of it.

In Japan, bugs are more than just a cultural phenomenon. They've become true status symbols. Stores in Tokyo are dedicated solely to selling live bugs along with all of their accoutrements. Finding a dealer can prove to be a bit trickier. They tend to be elusive due to the pricey treasure trove that they keep in stock.

A butterfly dealer in Tokyo is one such man. Though well known, he still manages to keep a low profile. His office consists of twin three-room apartments lined floor to ceiling with a filing system of slim maroon boxes. Each is filled to the brim with butterflies and moths, many of which are Appendix II species. Twenty thousand additional insects are added to his stash every month, having been gathered from hard-to-get-to places such as Papua New Guinea and Myanmar. All this comprises only a third of what he actually has in stock. The rest are kept under lock and key in a nearby warehouse.

He recently planned to travel to Tibet in search of a butterfly as elusive as Shangri-la, mainly because it's believed to be nearly extinct. The dealer will train and pay locals to continue the quest once he leaves. The butterfly will be worth a fortune if it's ever found. The man has already been jailed once in India.

Still other dealers have been run out of villages in Colombia and bushwhacked by natives in Irian Jaya. They take their lives in their hands on their butterfly-collecting trips. Everything they come up against has either a fang or a thorn in it. If not that, then they have bows and arrows.

The voracious demand has given rise to everything from hair-raising helicopter rides in Russia to ensnare mountaintop butterflies to poaching gangs that roam Central Asia. In each case, the butterfly's location remains a tightly guarded commercial secret. At times warlords and drug smugglers are involved in the mix. There's even a woman who controls all the street vendors selling bugs and butterflies in Thailand.

"She's the mama-san, and you do not mess with her," reveals collector and dealer Ken Denton, professionally known as "the Butterfly Man." "If you go into the wrong areas over there and try to open a butterfly stand, they trash your product and run you out of business." You know you've reached her abode when you arrive at a gated palace with satellite dishes and a battalion of Mercedeses parked in front along with a troop of armed guards.

"There are people who can get bugs out of any region of the world, out of any national park, out of any forbidden valley," confirms Brent Karner, associate manager of Entomological Exhibits at the Natural History Museum of Los Angeles County. "If you have the money, and want it, you can get it tomorrow. 'The Insect mafia' is certainly a good term for them."

Kojima also collected bugs as a child. But unlike the other little boys, he didn't keep them as pets. Instead, he immediately killed them. That way they remained perfect specimens.

Young Yoshi was free to indulge in whatever hobbies he liked, since he came from a family of means. The Kojimas prepared *kintoki* beans in the back of their home each day and sold them to restaurants and for bento boxes. The business flourished, and it's the same large house in which Kojima lives today.

His father was also a well-known butterfly collector and taught Yoshi everything he knew. The boy proved to be an apt pupil, and his reputation quickly grew. He was invited to attend a prominent Kyoto high school famous for its entomology club. He became the star pupil, only to clash with other students after a year and promptly drop out. From the beginning, he never played well with others. He then moved on to his next passion. He transferred to northern Japan and took skiing lessons.

Kojima participated in high-speed slalom races while attending Osaka University. He proved nearly good enough to make the 1971 Japanese Olympic ski team as an alternate. But he had to pay part of his way. He did that by working part-time at his uncle's antiques shop during high school and college.

Kojima bragged of having met his future father-in-law there, a rich American with a passion for Japanese antiques. His family had founded a big oil company that also happened to own a few banks. Charles Hanson soon began to make frequent trips to Kyoto to furnish his homes but mainly to spend time with Kojima, who quickly learned to enjoy the high life.

Kojima graduated from university with an entomology degree and, true to form, decided he no longer wanted to ski. He dropped the sport the day he left college. But then, it would have been difficult to find ski slopes in his new Hawaiian home.

Kojima had befriended a Mormon elder serving in Kyoto and, for a while, even became a Mormon himself. The man was then assigned to Hawaii for gospel work and invited Kojima to fly over for a visit. The plan was that Kojima would attend a summer course at the University of Hawaii to learn English. He packed his belongings, waved *sayonara*, and took off for Oahu, though he never bothered to attend the university. Instead, he quickly landed a job through a Japanese friend at a convenience store in Waikiki. The pay was lousy but Kojima soon found a way to remedy that.

A few representatives of the *yakuza* came into the shop one day and approached Kojima for assistance. They'd recently expanded their business and had begun making movies in Oahu. The problem was the subject matter. The films they made were more *Yuki Does Tokyo* than *Bambi*. That was fine and dandy in the United States, but fully visible hard-core action pornography is taboo in Japan. What was a group of enterprising new filmmakers to do? They knew their eight-millimeter movies would be a big hit on the Japanese black market. The trick was how to sneak their films into the country. Could Kojima possibly help them? They'd gladly pay for his time and trouble.

Being a natural entrepreneur, Kojima pondered it for a while and came up with an ingenious plan.

The shop sold large boxes of chocolate-covered macadamia nuts. It didn't take Kojima long to realize how else they could be used. He carefully opened one end of the cellophane wrapper, removed the box, and ate all of its contents. Then he cleverly replaced the candy with two of the movies and slid the box back into its transparent shroud. The weight of the film proved to be almost exactly the same as that of the nuts. Kojima seamlessly ironed the cellophane closed, and voilà! The problem for organized crime had been solved.

The *yakuza* couldn't have been happier with his scheme as they proceeded to rack up air miles between Hawaii and Japan. They always transported filled boxes of chocolate-covered macadamia nuts inside their shopping bags, and every trip proved to be lucrative. The films sold on the Japanese black market for a thousand dollars a pop. The plan worked for three years, until Customs finally caught on that not all the *yakuza* in Japan could have that much of a sweet tooth. By then, Kojima had not only learned the joys of smuggling, he'd bought a seventy-thousand-dollar condo in Oahu with his share of the proceeds. He claimed to have sold it a few years later for one million dollars. Kojima was Donald Trump and Wile E. Coyote rolled into one.

He was flying high enough to visit his family in Kyoto on a regular basis. One trip found him seated next to the vice president of Pepsi Japan. The VP must have recognized Kojima's unique talent and initiative. By the time the plane landed, Kojima had been offered a job. He was invited to manage Pepsi Japan's new discount travel agency for its employees.

Kojima moved back to Japan, where he learned the business from the ground up. The agency soon outgrew its one-room office and in almost no time occupied an entire floor. But Kojima was never satisfied with doing just one thing for very long. In addition, he began selling antiques to foreign dignitaries out of his apartment in the Roppongi district of Tokyo.

Roppongi is an area infamous for its risqué shops, strip joints, dance halls, and sex clubs. It's the place to go to enjoy the kinkiest in adult toys, porn magazines and videos, and S&M gear. Kojima spent his evenings dabbling in all that Roppongi had to offer, as well as entertaining beautiful European models both male and female. But he always reserved vacations to be spent with his future father-in-law.

The two men took excursions together and experienced the pleasures of Italy, Spain, Germany, and France. Kojima was even flown to his father-in-law's house and invited to live with him there. He stayed for two months, hated the place, packed his bags, and returned to Japan. The next proposal was that the men reside together in Los Angeles. Kojima agreed but only if he could start his own discount travel agency specifically targeted at Japanese travelers. His father-in-law helped make the arrangements, and Jet Sky Travel was born. Kojima would later claim that his business had a branch office in Tokyo with twenty employees and another in Hong Kong with seventy associates, as well as the main office in Los Angeles, where six hired hands worked for him.

Kojima was on a roll and life couldn't have been better.

Inside a Smuggler's Lair

Obsession is medicinal poison.
— Christopher Molineux

Kojima wasn't about to let any grass grow under his new protégé's feet. He sent Ted Nelson an e-mail immediately after their Starbucks meeting. Would Ted look at a Web site called InsectNet.com that also auctioned butterflies? It might give him some ideas. Kojima followed that up with a phone call later that evening. Once again, it was after hours and Newcomer wasn't at his office to answer the undercover phone. Kojima left a message that Nelson should return his call ASAP.

Newcomer was learning the joys of having a government job. Agents are expected to handle as many cases as they can juggle, along with all the paperwork and a continual backlog of reports to be filed. Papers had already begun to breed rapidly on Newcomer's desk, where they were well on their way to mutating. One wrong move and the piles would come tumbling down. Picking up the phone and placing a call was about the last thing that Newcomer wanted to do at the moment. Besides, that's what e-mail was for. He dashed off a quick note to Kojima.

I'll take a look at the site. Is InsectNet's setup the sort of thing you have in mind? I have a meeting in Long Beach this afternoon but can speak with you tonight or anytime tomorrow.

Kojima promptly read the e-mail. Nelson was crazy if he thought that Kojima would be kept waiting. He immediately picked up the phone and called.

Newcomer just missed knocking over a jumbo cup of Coke and a bag of Doritos as he lunged for the phone.

"Ted, this is Yoshi."

No fooling. Maybe it was Kojima's tone of voice, but Newcomer felt like a schoolboy who was about to be scolded. "Hey, Yoshi. What's up?"

"We need to meet again for another few hours. There's still much to discuss about eBay. I want to start our new business as soon as we possibly can."

Newcomer was anxious to do it too, but didn't this guy ever get tired of talking? They'd just hashed things out for over an hour at Starbucks. How much more could Kojima possibly have to say on the topic? Especially since he supposedly didn't know anything about eBay.

"Besides, I may have some butterflies for you before I leave on my Costa Rica trip."

That did the trick. Newcomer agreed to meet outside Starbucks on Friday and go to lunch from there.

"And don't forget to do research on how eBay works before then," Kojima reminded him, cracking the whip.

Terrific. Newcomer now had two taskmasters, and Kojima was proving to be just as demanding as the government.

He stuck his head inside Palladini's office. "Hey, boss. I've got another date lined up with Kojima. Do you think I could get an agent to provide backup?"

What the heck. It was worth a shot. After all, *Adam-12* had partnered Reed and Malloy, and then there were Starsky and Hutch. It seemed like the right thing to do. Just by luck, San Diego agent John Brooks planned to be in L.A. that day. Brooks would follow Newcomer from Starbucks to lunch, where he'd wait outside in his car.

"Great. I'll signal you if there's a problem," Newcomer told him.

Brooks offered in turn to call during lunch and make sure that everything was cool.

"Other agencies think we're nuts for some of the stuff we do," explained Brooks, a twenty-six-year Fish and Wildlife veteran. "We go in by ourselves, no backup, no contact, no weapons, no nothing."

The Service is notoriously hampered by a lack of manpower and money, with fewer than two hundred special agents in the country. Those working cases tend to play the Lone Ranger, minus Tonto and Silver. Brooks had been in plenty of crazy situations himself and knew that having backup was rare. That makes it a deadly game of Russian roulette when it comes to dealing with poachers and smugglers. Fish and Wildlife agents had just been plain dumb lucky so far. There'd come a time when their luck would run out.

Newcomer pulled up to Starbucks on Friday to find Kojima already waiting for him. The guy could have been a transplant from Miami dressed in his usual khaki shorts with sneakers and white socks. Kojima hurried over and hopped into Newcomer's car.

"It's easier if we go together. I'll show you the way. Just keep driving down Venice Boulevard," Kojima said, while his fingers fiddled with something in his fanny pack.

Newcomer just hoped he wasn't about to pull out some butt-ugly bug. He glanced back stealthily to make sure that Brooks was still there.

"Look at this e-mail I just got," Kojima bragged, and began to wave a piece of paper in Newcomer's face.

Newcomer perused it while waiting at a series of stoplights. Holy shit. Kojima had a laundry list of insects that had been recently gathered for him in Central America. National Geographic was sending him to Costa Rica the next week, and he planned to pick them up while he was there.

Hercules and horned beetles were at the top of the list, along with scarabs in a rainbow of colors. The bugs were itemized in tempting hues of silver, semi-pinks, reds, and rare gold as though they were tantalizing shades of nail polish. There were also boucardis and tapantinas, bugs that he'd never heard of. He couldn't quite believe what he read. The grand tally came to seventy-nine insects for a whopping total of only $225. It was Kojima's version of shopping at Walmart.

"Oh, I'm so happy! Some of these will sell for between three to five thousand dollars apiece," Kojima gloated.

Newcomer's fingers tightly gripped the paper, not wanting to let go. This was his first glimpse into Kojima's global network and the bargain-basement prices that he paid for insects. "Wow! So who's doing all this collecting for you?" he asked, the paper beginning to feel moist in his hand.

"Oh, that's from my Mexican guy. He worked as a gardener here in L.A., but Immigration catch him and he got deported. That's okay. Now he works for me. He's such a nice boy." Kojima retrieved the note and played with his fingers as if counting all the money that he would make. "I send him all over Central America to get butterflies and bugs that I want. Then I go and pick them up. I pay him two thousand dollars a month, plus all expenses. It's very cheap."

No wonder he was the go-to guy for butterflies that nobody else could get. Instead of a girl in every port, Kojima had a bug catcher in every corner of the world.

They pulled up to Natalee Thai Restaurant, and Kojima stepped out of the car. Newcomer was about to follow when he froze and caught his breath. Kojima's shopping list of bugs was lying on the seat. Kojima must have left it there by accident.

Was it possible? Could Kojima really be that forgetful?

Newcomer had a choice. He could swipe the list or prove himself to be a trustworthy associate. Screw that. As far as Ted Nelson was concerned, it must have somehow blown out the window. Newcomer snatched the paper and slipped it into his pocket, snagging a piece of evidence.

Kojima launched into business as soon as they sat down at the table. "Did you get an eBay account yet?" he prodded.

Man, this guy was like a pit bull. They'd just begun doing business together, and already he didn't let up. Newcomer shook his head. "Nope. I plan to do it while you're out of the country. Don't worry. It will be ready to go when you return."

Kojima studied Newcomer as if sizing him up. Perhaps the boy wasn't as confident as he pretended to be. "If you're afraid, we can start off using my former company name, Butterfly and Insect World. But to be honest, using 'Ted Nelson' would really be better."

I'll bet, Newcomer thought. Kojima was clever as a fox.

"So, exactly how is this going to work?" Newcomer asked.

"I'll supply all the butterflies. You collect the money and ship the merchandise to the customers. Then we'll split the profits fifty-fifty," Kojima explained.

Yeah, yeah, yeah. Newcomer was still waiting to hear the catch. "What about permits? Will there be any problems that I should know of?"

"I don't have permits, but there's nothing to worry about. You won't be reported. Most people don't even know to do that. Besides, no one really cares about getting permission for butterflies," Kojima assured him with a wave of his hand.

That was interesting to know, but something else caught Newcomer's attention as his eyes locked on Kojima's hands. They looked like they hadn't been washed in a week; a strip of dirt lay buried beneath each fingernail like an animal hibernating for winter. *Oh, Christ. Just let me get through lunch,* he silently groaned.

"Listen, I don't mind selling exotic butterflies. I just don't want to get caught, is all."

Kojima irritably shook his head. "Why are you so scared? I tell you there's no problem. I send all my packages Express Mail from Japan. The Customs people never bother to check." Kojima had been relishing his food, but his appetite now started to wane. "You worry too much. I already explained. We'll start with common butterflies. As you work, you'll get to know the customers. Then we can sell CITES material to them. That's where the real money is."

Perfect. He'd gotten Kojima conspiring on tape. Now all he had to do was to catch him in the act.

"That's great, Yoshi," Newcomer agreed. "As long as you're sure that everything will be okay, then I'm absolutely fine with it."

Kojima's appetite perked right back up. "I can't wait to get started! I sold a CITES II butterfly for eighty dollars to a guy that turned around and sold it on eBay for five hundred bucks a few weeks ago," Kojima revealed.

He cautioned there might be one problem, however, once they began offering CITES material.

Wasn't there always something? If smuggling was that easy, everyone would be doing it. "What's that?" Newcomer asked.

"All birdwing butterflies are listed as Appendix II. Some people might ask questions and want CITES permits from us. If so, you must immediately drop them like hotcakes," he warned in his thick Japanese accent.

"Why? What can they do about it?" Newcomer asked, extracting information bit by bit.

"They might be bad people that want to report us to Fish and Wild. You let me know if they bid on any butterflies, and I'll outbid them using a different name. That way they'll have no proof. Also, you must keep a list of their names so that we never sell to them," Kojima advised.

"Yeah, but what if they ask where I get the butterflies from? Should I say they come from you?" Newcomer inquired.

"No, absolutely not! Just say that you get them direct. Don't ever use my name," Kojima admonished.

Wasn't Kojima the careful guy? This was becoming more and more interesting. "How come? Because they'll think you got the butterflies illegally?" Newcomer asked, fishing carefully.

"Yes. Another thing you should know is how to get around Fish and Wild if they contact you. Just say your boss has all the permits in Japan and that you only ship things out. But don't worry. Only once did they send their very top guy to my house." Kojima grinned, as if relishing the memory. "It was so funny. This guy came with a gun and handcuffs on his belt and tried to scare me. He didn't have a warrant, but I let him in anyway because I never keep illegal material in my L.A. home. That's because I'm smart."

Keep bragging, Yoshi. We'll see just how smart you are.

Newcomer was feeling like the cat that was about to eat the canary. He figured that lunch was over and it was time to get rolling. Only Kojima wasn't quite finished with him yet.

"Let's go to my apartment. Maybe I have some mounted butterflies that you can put up on eBay while I'm away," he suggested.

How cool is this? I'm now being invited to the world's most notorious butterfly smuggler's place! Newcomer could barely contain a laugh.

The stakes had just been ratcheted up. Kojima clearly trusted Ted Nelson. It was unexpected, but he and Brooks should be able to roll with it. Besides, no way was he about to let this opportunity drop. After all, how many people got to go inside a smuggler's lair?

"That sounds great, Yoshi. Just give me a minute to hit the head," Newcomer said, and excused himself.

He dashed to the bathroom, locked the door, pulled out his cell phone, and gave Brooks a call.

"Hey, there's been a change of plan. I'm going to Yoshi's place. Be ready to follow. We're coming out in a minute," Newcomer whispered.

He barely heard Brooks's response over the beating of his heart as someone started to turn the knob. Newcomer slipped his cell phone back in its holder and went out to join Kojima to discover he'd already paid the bill.

"Okay, let's hit the road," he said, and the two men left the restaurant and got into his car. Pulling out, he made a sharp left onto a busy boulevard.

So far, so good, Newcomer thought, excited to be making headway so quickly with Kojima. He was in for another surprise as Kojima pulled a CD-ROM from his fanny pack. The pouch was proving to be a never-ending treasure trove. It was nearly as good as a magician's hat. Newcomer never knew what was going to come out of it.

"This is for you. It has all of my picture files with thousands of butterflies. You can use it to post photos on eBay of the butterflies that we offer," Kojima instructed.

Newcomer would soon learn that among the bugs on the disc were not only many protected species but also endangered ones.

Kojima was chatting away when Newcomer stopped at a light and glanced back in his rearview mirror. *Damn!* John Brooks's car was nowhere in sight. What in the hell had happened to him? His eyes frantically searched the road. He must have been swallowed by traffic or gotten stuck at a light. Either way, Newcomer was in trouble without backup. He did the only thing he could immediately think of. He slowed to a crawl, hoping to give Brooks a chance to catch up.

But it was as if Kojima could read his mind.

"You go too slow," he scolded, and then issued a command. "Turn here!"

Just great, Newcomer thought as he found himself entering an underground garage. *Brooks is never going to be able to find me here. Now what do I do?*

He quickly hit upon a plan. "Hey, I've got to call a buddy that I'm supposed to meet. I want to let him know I'll be a little late," he informed Kojima. He'd casually mention Kojima's address, and that would get Brooks there on the double. He gave himself a mental high-five as he stepped out of the Tahoe and reached for his cell phone only to have the damn thing pop out of its holder, bounce like a spastic gymnast, and land smack on the ground.

Hell! He tried to turn it on, but the phone was dead. Now he not only had no backup, he had no cell phone either. Could things get any worse?

Yep. There it was, that familiar sickening feeling. A wave of panic hit and paranoia wormed its way into his gut as Newcomer now began to have a conversation with himself. *Oh, come on. There's nothing to worry about. You can easily take this guy. Except there's always a chance that someone else could be in the apartment. And what if Kojima has a gun?* Kojima had checked Newcomer's car as thoroughly as if he planned to buy it, and Newcomer had no idea whether he might have figured anything out. He didn't trust Kojima for a New York minute. This could very well be a setup.

What Newcomer did know was that the situation was dangerous and he had to think fast.

"Would you look at this? The phone's just a lousy piece of plastic," he said, showing it to Kojima while stalling as long as he could.

"The next time you should buy Japanese," Kojima replied with a laugh, walking toward the building's stairwell. Newcomer was running out of time and had to quickly decide what to do.

He could always pretend to be hit by a sudden attack of stomach cramps and postpone the visit. Or he could suck it up like his heroes Reed and Malloy, G.I. Joe, and Kwai Chang Caine. God knows, he'd learned as a boy how to instantly invent a story, come up with a good excuse, and spout just the right line. He'd bluffed his way through plenty of situations before.

Newcomer made up his mind. He was bound and determined to get inside Kojima's apartment. He wasn't about to back down now, but it would have to be quick. He probably had twenty minutes tops before things could get ugly. Brooks would most likely call the LAPD if he couldn't find Newcomer or reach him by phone. Then the police would come looking for him and kick down Kojima's door. Once that happened, Newcomer's cover would be blown and the case would be game, match, and over.

He glanced at his watch and marked the time. *I guess this is what I signed up for.*

He followed Kojima inside the white stucco building.

Kojima opened the apartment door, and the stench that raced out took Newcomer by surprise. Good Lord, what was that smell? It was nearly palpable, as if something had died inside and never been removed. A shudder ran down Newcomer's spine as the scent wrapped itself around him tight as a shroud.

He took a step inside the dark apartment, and his skin began to crawl. There was something in this place that definitely creeped him out. Another step and whatever was sour now took root in his stomach. There was no escaping it. The putrid odor most likely permeated the living room, the kitchen, the bathroom, and hall. It contaminated the air and seemed to cling to the walls like an invisible veil of fog. Then it dawned on him. It was the reek of insects. The place had to be filled with bugs, both dead and alive. It was Newcomer's worst nightmare come true.

He took a quick look around. What in the hell was this place? The movie set for *Psycho*? The living room was a cluttered jumble of junk, and the soiled curtains were closed. Maybe it was a good thing. At least that way he couldn't see all the dirt. What he did see was that almost every available spot was piled high with Tupperware containers and wooden boxes for transporting butterflies. The remaining space was littered with an assortment of empty fast-food cartons. As for the floor, it was fertile ground for a mountain of Japanese newspapers and magazines.

A rumpled blanket partially covered a shabby old couch whose fabric was a pop-art array of stains. One cushion was gently caved in like a fallen soufflé. It must have been where Kojima sat and watched

TV. The only items out of place were the antique Japanese screens lining the walls. There were also a few beautiful old vases hidden like gems among all the crap.

Newcomer's fingers crept to the back of his neck, where he was certain a bug had begun to crawl on him. Nope, nothing there. It was the feeling that he was being watched. He craned his neck and peered down the hall. He could see into the bathroom, where a black ring of grime encircled the tub. A layer of dust and dirt seemed to cover everything. Even the doorknobs looked filthy. There was probably a bedroom, but no way was he about to go explore. He had no backup, no phone, and no gun. It wasn't worth the risk just to satisfy his curiosity. The wiser choice was to stay put by the front door, when he realized that Kojima had disappeared.

"Come here. I want to show you something," Kojima called out, sounding excited as a child.

Newcomer followed his voice into a room that was more a science experiment gone wrong than a kitchen. There were no dishes, pots, or pans in sight. Instead, plastic containers were haphazardly stashed everywhere. They were precariously balanced in the cupboards and on the counter. Kojima opened a decrepit refrigerator and containers filled all the shelves. What made it particularly macabre was that bugs were entombed inside. Some were trying to climb out of their plastic mausoleums, while others had simply given up and died.

Kojima reached for one box, his fingers seeming to transform into ten wriggling worms as they eagerly lifted the lid and slid down into the dirt. Watching him poke among what looked like dozens of slithering grubs, Newcomer now knew why Kojima's fingernails were black.

"These are *Dynastes*. They're building their cocoons. When I return from Costa Rica they'll be ready to hatch. Then I can kill them and take them back to Japan to sell," he said, his voice gurgling with delight. "I'm the only person in the world who can successfully breed them because I know the trick."

Kojima placed the box back in the kitchen cupboard next to hundreds of other bugs waiting to cocoon. Then he grabbed a container that held a dead black beetle the size of his fist.

"This is *Dynastes hercules,*" Kojima nearly crooned while prodding it with his fingernail. "They fly at midnight, so I have to shine a bright light to catch them. They don't bite, but they grab on to my hands and it's very painful."

The sharp stench of mothballs seared Newcomer's nose as Kojima now tried to hand the bug to him. Was he out of his ever-loving mind? No way was he about to touch that thing. *I've got to get out of here now!* Every fiber in Newcomer seemed to scream as Kojima brought the bug closer. Newcomer's arms remained plastered by his side, and he did his best not to flinch. Enough was enough. It was time to get down to business. "So, do you have some butterflies for me?"

Kojima put the bug down and walked into the living room, where he began to rummage through a pile of boxes.

Phew! That was close.

"Sorry everything is so messy," Kojima mumbled. "But I'm packing to move back to Japan."

"You're going back to Japan?" Newcomer asked, caught by surprise. "But what about our business deal?"

Kojima thought it was sweet. The boy actually seemed to care about them working together.

"It's no problem. The Internet is the future for selling butterflies. Business will be very good," Kojima assured him. "Besides, I'll still keep an apartment in L.A., but not this one. Oh, look what I just found."

He pulled an eight-by-ten photo from a cardboard box. It was a picture of himself dressed in a slalom suit skiing through a gate decorated with little Olympic symbols.

"I used to be a very good skier when I was young. In fact, I was an alternate on the Olympic team in Sapporo," he boasted.

Kojima next produced a head shot of a well-known TV actor.

"And here's a photo of my very good friend Sonny. We met at Sports Club/LA in Beverly Hills. I have this apartment for another two months. After that, I might move into one that he owns in Marina del Rey. It's much nicer. There are some very bad people in this building. I've had a lot of trouble with one guy in particular. This is not a nice place."

Kojima was referring to his neighbor and former friend, Alexander Denk.

DENK IS A BODYBUILDER who has won the titles Mr. Vienna-Austria, Mr. Natural Universe, Mr. World Musclemania, and Mr. Natural Super Bowl Champion, among scores of others. His résumé of skills reads like a laundry list. He's a personal trainer, a professional chef, a nutrition expert, and a real estate investor, along with being an actor and a stuntman. He's actually been Arnold Schwarzenegger's body double in five films. Denk even looks a bit like the Austrian-born bodybuilder California governor. He calls himself "the Denkster" and has an old video running on YouTube. He once wanted to be the next reality fitness star.

His biggest claim to fame, however, is having worked on Anna Nicole Smith's reality show. He went from being her chef, to her personal bodyguard, to her lover. He announced after her death that there was a good chance he was also the father of her baby.

Kojima and Denk met at a popular breakfast spot when Kojima still had his travel agency. He snagged a discount airline ticket for Denk, and a friendship ensued. According to Kojima, Denk was usually short of cash, and Yoshi frequently paid for his meals. Kojima lived in Mount Olympus at the time and would hire Denk to watch his pets whenever he went away. He once told Denk that he had to leave on a trip because National Geographic was going to do a story on him.

"Yoshi had three or four dogs in cages for months on end and there was shit all over the place. One dog was all white, and the other was a pure black German shepherd. The poor things would go crazy whenever he'd finally let them out. He didn't treat them right," Denk complained.

But what amazed him most were all the bugs in Kojima's place. There were beetles in containers that took up every square inch of the kitchen. It was one reason why Kojima never ate at home. The other was because none of the cooking appliances worked.

"He also had two parrots and betta fish in amazing colors that no one had ever seen. The guy made money at whatever he touched. But the house was a total pigsty. In truth, I don't know how he could live that way," Denk confided.

What now looked like a slum had once been a beautifully maintained luxury home with a million-dollar view, a large pool, a black stone floor, and expensive Japanese antiques. That was when Charles Hanson used to visit. Denk had seen photos of the two together and had even met the tall, elegant man with a shock of white hair a few times. Kojima and his father-in-law were no longer as close but still remained good friends. The only thing Denk didn't know about was Kojima's wife and son.

Except for when he was with Denk, Kojima now spent all his free time at home alone mounting butterflies

"For me, he was a genius, this guy. Yoshi had a brilliant mind. He gave me butterflies and a big Sony TV," Denk reminisced.

But there was another side to Kojima, which came out whenever he felt he'd been double-crossed.

Denk became determined to learn everything there was to know about the insect business, and for that there was no better teacher than Kojima. He was especially fascinated by the thousands of butterflies that Kojima kept stashed in his walk-in closet. Collector's wooden boxes filled shelves that stretched all the way up to the ceiling. There were even a few Queen Alexandras, an endangered species, among the lot. Another closet held hundreds of plastic containers with still more butterflies piled high, their wings folded as if they were sleeping.

The two men began to travel together to Payson, Arizona, to collect Grant's rhinoceros beetles. People were always amazed to see the Japanese dealer chasing after bugs with the large muscle-bound Austrian bodybuilder by his side. They looked like a modern-day version of Mutt and Jeff. The two would gather eight hundred bugs at a time and wipe the area clean. However, the men eventually had a falling-out, over money.

Kojima's travel agency took a nosedive after 9/11. He soon began dodging not only creditors but also the state of California. He hadn't paid state income tax for the previous four years, and the amount due had climbed to $42,656.33. Kojima had managed to cleverly elude the tax collector so far by listing his address as what turned out to be a drop box. But some creditors had caught a whiff of his scent and were beginning to close in on him.

Kojima decided it was time to put the house up for sale, get what money he could, and vamoose. It just so happened that Denk knew the right buyer. He also thought it was understood that Kojima would pay him a commission for his part.

Kojima received six hundred thousand dollars. But his problems weren't over just yet. He now had nowhere to live. It was impossible to rent a place without a credit check, and Kojima was still on the run from creditors. Denk flexed his muscles and once more came to Kojima's aid. This time he pulled a few strings with the owner of his apartment complex.

There was just one hitch. Kojima explained that the apartment would have to be in Denk's name, along with the phone and utilities. After all, he was still dodging creditors. Denk had yet to receive the commission he felt was due him and had little choice but to reluctantly agree.

Kojima proceeded to ditch the two parrots, the remaining fish, and most of his dogs. He moved in, bringing with him a couple of thousand bugs and butterflies, one German shepherd, and all his antiques.

Five months later, Kojima filed a police report claiming that some of his things had started to vanish. He suspected the culprit was Denk.

An investigation was launched and police interviewed witnesses who denied Kojima's account and questioned his credibility. As a result, Denk was never charged. The two men haven't spoken since.

"I THINK I MIGHT HAVE FOUND something for you," Kojima said, and opened a wooden box.

Newcomer gazed at each luminous birdwing butterfly inside, finding it hard to believe they were real. They looked like multicolored rainbows fallen from the sky, their wings a fantasia of Technicolor glory.

"Oh, no. I'm so sorry. These are no good after all," Kojima said abruptly in what seemed a deliberate tease. He closed the box and quickly put it away. "I must have already sent the ones that I was thinking of back home to Japan."

What? Newcomer looked at him in disbelief. Kojima had to be kidding. Wasn't this why he had brought him to the apartment in the first place? "That's okay, Yoshi. They look fine. We can start with these," Newcomer said, itching to get his hands on them and leave. Time was ticking away.

"No, no. They don't have the proper species identification you need for writing descriptions. Besides, these are not perfect. We can only sell flawless specimens on eBay, otherwise customers will complain. Don't worry. I'll bring new ones back for you from my trip," he promised.

Newcomer snuck a look at his watch. His hourglass had officially run out. He had to leave now or risk blowing the case.

"Then I guess it will just have to wait," he reluctantly conceded, and began to inch his way toward the door. "Listen, I have to meet my friend, so I'm afraid I have to go."

"Really? You need to leave so soon?" Kojima asked, clearly disappointed. "I guess you probably also have a girlfriend to see, right?"

Newcomer stood still for a moment as every nerve in his body strained. That wasn't a siren in the distance, was it? "Sure. I have lots of girlfriends, Yoshi. I like to play the field." *Keep it vague,* Newcomer reminded himself.

"That's good," Kojima mumbled glumly, appearing to be momentarily distracted. Then he snapped back to attention. "I think we should meet once more before I leave for Mexico and Costa Rica. We still have some things to discuss."

Sure, whatever. Just get me out of here before the troops arrive. "That sounds great, Yoshi. Why don't you give me a call?" he suggested while reaching for the door.

Newcomer's palm itched as soon as it touched the knob. The surface felt gritty, and his nerves were about to explode. He beat a hasty retreat out the door, ran down the stairs, jumped into his Tahoe, and raced to the nearest pay phone. He had to reach Palladini before what had turned into a comedy of errors ended with a police visit.

His words flew out in a rush when Palladini got on the line. "There was a screwup! Brooks lost me in traffic and my cell phone broke," he said in one breath

Palladini took it in stride. "It's okay. We're in the middle of L.A. I knew you could handle yourself and that everything would be fine."

It was Brooks who chewed him out. "Hey, you were only supposed to have lunch with Kojima. Going to his place wasn't part of the plan. What the hell happened to you after that?"

"What do you mean, what happened to *me*? What happened to *you*?" Newcomer shot back.

He discovered that things had quickly gone awry after Brooks hit a stoplight.

"By the time it turned green, you were gone. Worst of all, you didn't answer your damn cell phone! What's wrong with you? I had no idea what was going on," Brooks exploded.

He'd freaked out while trying to locate Newcomer's Tahoe for the next thirty minutes. There was no way for him to know the vehicle was safely tucked away in an underground garage. Making matters worse, Brooks had attempted to do a one-man surveillance in unfamiliar territory. He'd called Palladini for thc address, but Kojima's former Mount Olympus home was all that was listed on his license. There'd been no choice but to let the situation play out.

Things slid downhill from there. Not only did Brooks get lost in traffic and Newcomer's cell phone break, but Newcomer then learned that his digital recorder had accidentally shut itself off. There was no official record of the meeting. It was Newcomer's initiation into the dangerous world of undercover work.

He drove home with the stench of Kojima's apartment filling his nose and clinging to his clothes like stale cigarette smoke. He raced into his place and jumped into the shower, but no amount of scrubbing could seem to take the smell away. Nor could he erase the image of those bugs desperately trying to climb out of their containers. It made no difference that he'd never touched them. The sensation was as if he had bugs crawling all over him.

He went to sleep that night still hearing their legs scratching futilely against the plastic until the insects slowly cocooned and finally metamorphosed. Only then could Newcomer finally go to sleep. He dreamed of fields of dead butterflies, rigid as tin soldiers, all perfectly lined up in rows.

Growing Up MacGyver

The work is a calling. It demands that type of obsession.
—John Pomfret

Ed Newcomer was always small for his age, which can be hard on a kid. He grew up on the wrong side of the tracks in Denver and became an easy target for bullies. He was five years old when his dad walked out. His mom and two sisters were left to raise him. It was 1970, the days when it was difficult for an unmarried woman to get a credit card, much less write a check. His mom worked days as a legal secretary and nights at a second law firm to keep their family afloat. Newcomer was keenly aware of how tight money was and how hard she struggled just to buy necessities. His mother became the queen of jerry-rigging, and he learned the skill from her. That knowledge would impact him for the rest of his life and turn Newcomer into the MacGyver of the U.S. Fish and Wildlife Service.

He was always amazed at how she created toys out of the simplest of things. To Newcomer, it seemed like magic. He once wanted a lasso for a prized cowboy doll. His mom showed him how to take a string, neatly wrap it around three fingers, and tie it off. It not only looked like a lasso but also unrolled like one. Cardboard boxes morphed into houses to shelter his action figures. The best thing was he could always build a new structure if it broke.

Young Ed was a huge fan of Bigfoot. The creature represented everything that was unknown and scary about the world to him as a boy. He'd always have one of his G.I. Joes searching for the creature as he strung complicated adventures together for weeks on end. The careful planning that went into it would prove useful later when it came to undercover work. He now views Bigfoot as the poster child for endangered species.

Ed also imagined people breaking into his home and concocted elaborate schemes to save his family. His fantasies involved devising intricate traps and high-kicking karate fights. He'd painstakingly work out each detail, driving his sisters crazy. But he was deadly serious about his plans. "I know what to do and how to protect us, so just be ready," he'd warn.

Most of all, he enjoyed pretending to fight for those who couldn't protect themselves, be they animal or human. In truth, Newcomer was the one who needed help. He was constantly picked on because of his size. Other kids instinctively sensed his anxiety the way that animals smell fear.

Ed attended a Denver inner-city school and quickly learned that things could get rough. Inventing stories was his best way of escaping trouble, and he became a master in the art of fibbing as a means of survival. Even so, he was the perfect bull's-eye for bullies.

He was pushed against walls, punched, and choked; once he had a screwdriver held to his neck. He walked the halls in fear of what lurked around the next corner. Things were so bad he avoided the bathroom all throughout junior high. It was the most dangerous territory in the entire school system.

His biggest fear was that someone would take his bike and he wouldn't be able to stop him. Luxuries didn't come easily, and he couldn't bear the thought of losing a beloved possession. He hated school and was plagued by insecurities. His only escape was pretending to be an action hero. All he wanted to do was grow up.

Ed finally decided enough was enough. He was tired of being a target. It was time to take a stand. He religiously watched the TV show *Kung Fu* and dreamt of becoming David Carradine's character. He begged to be allowed to take a martial arts class.

As a Mennonite, his mother was against violence, and it proved to be a hard sell until the day Newcomer walked in after having been beaten up. She promptly went to register him for a summer course in tae kwon do. The YMCA class came complete with torn wrestling mats, dusty curtains, and sometimes no heat in the winter. What it did have was a master tae kwon do instructor. The catch was that the teacher didn't take children. Try telling that to a woman on a mission. Ed's mom talked him into giving her son a shot. If he didn't keep up with the class, then the teacher could simply boot him out.

Ed was instantly hooked. Once he began, he didn't want to stop. By the ninth grade, he had earned a black belt and a lot more confidence, and kids had stopped picking on him. The worst that would happen was the occasional bump in the hallway. He went on to train with his instructor for the next ten years.

All his childhood hardships eventually played to his advantage. Newcomer proved to be a natural when it came to undercover work. He could bluff his way through almost anything and had learned to roll with the punches. His adolescent role-playing provided the basis for his career, while tae kwon do training made him a tough adversary.

The best undercover agents are always ten steps ahead of their prey, using strategy much like in a chess game. In addition, successful undercover work depends on three fundamentals: remembering what you tell your target, remaining as vague as possible, and all the while continuing to act naturally. Forget one of those rules and your case could quickly fall apart. Newcomer had all of these elements in place.

Having the know-how of MacGyver was also a plus. Fish and Wildlife tends to work on a shoestring, and sophisticated equipment isn't usually available. Agents will cobble together whatever's needed for undercover work, building their own covert cameras or creating garments with secret pockets. The job requires creativity, ingenuity, and tremendous perseverance, as well as comfort in working alone.

Since animals can't talk, many times agents are the only witness to a crime. That makes undercover work essential when it comes to cracking wildlife cases.

Newcomer had one more thing going for him. He looked earnest enough to be an altar boy. It made him all the more lethal. What he didn't yet know was how addictive undercover work could turn out to be. For most who took on the job, it becomes an obsession.

The Good, the Bad, and the Ugly

The caterpillar does all the work,
but the butterfly gets all the publicity.
—George Carlin

Butterflies are beautiful, ethereal creatures. They're fluttering tapestries of color, magically moving canvases, and tiny aerial dancers. We revere them in the form of tattoos, celebrate their shape in jewelry, and adorn ourselves with butterfly hairclips and T-shirts. Paper butterflies even festoon brothel rooms in India, as if to help to make the girls' lives a bit more palatable.

Butterflies are everything we seemingly long to be, young and exquisite forever. They're nature's supreme work of art and had become Kojima's sole obsession.

"Butterflies are the one bug in our society that people like because of the character traits we put on them. They're colorful and pretty, they're angelic and nice, and that makes them lovable," explains Brent Karner of the Natural History Museum of Los Angeles County.

However, the real life of butterflies isn't all it's cracked up to be. They're beset by creepy predators, eaten by birds, chased by humans, squashed on windshields of cars and caught in their grilles. Only 2 percent of the eggs laid by a female butterfly ever make it to adulthood. Perhaps because of this, many of them also have a dark side, along with rap sheets that can seem unbelievable.

Butterfly society is filled with various forms of brutality, from cannibalism and cyanide poisoning to molestation of minors and outright rape. They have more in common with humans than we might like to admit. Their actions can read like a tragic Shakespearean play. They're Coriolanus, Titus Andronicus, and King Lear all rolled into one big chrysalis.

From the moment of birth butterflies display an aggressive side. Survival of the fittest is the motto of many of these young cannibals as comma-sized caterpillars munch away on their siblings. They furiously chew in the hope that they'll turn into pupae before they themselves are eaten by birds.

If the youngsters make it that far, a worse fate possibly awaits them—the dreaded ichneumon wasp. The wasp perches on the back of a defenseless caterpillar and lays its eggs beneath the skin. Their offspring eventually hatch and feast on their host from within. In a nightmarish scenario, newborn wasps, rather than a beautiful butterfly, ultimately emerge from the chrysalis.

Other parasitic wasps detect the odor of a newly mated female butterfly and hitch an uninvited ride to her egg-laying site. There they proceed to parasitize her freshly laid eggs. Only one one-thousandth of her eggs survive.

Perhaps strangest of all are butterfly-mating rituals. We tend to romanticize these delicate creatures, but some of them can be downright cruel, participating in an act that's nothing less than cradle rape. There are butterfly species so eager to find mates that they locate a female pupa and perch on it, sometimes four males at a time. Then they wait until the female emerges. That's when she's at her most susceptible. Her wings have yet to open, and she's still limp.

Facing abdomen to abdomen, the first male's claspers spring from his sides, exposing his penis. The claspers then grab hold of the female and pry open her genitalia. Once he's finished mating, the rest of the males take their turn. The last male to mount is the one that fertilizes her eggs. He claims his prize by depositing a sperm plug, the butterfly equivalent of a chastity belt. That female is then unable to mate with anyone else. She may even be injured during the rape and die.

The *Heliconius* butterfly of Central and South America elegantly sports long, narrow black wings and dramatic yellow zebra-like stripes. The species is intelligent, with a brain twice the normal size of other butterflies. Large groups of them roost together on tree limbs at night. There they make an eerie creaking sound by wriggling their bodies to scare away predators—not that they'd have any problem. Their wings are loaded with cyanide.

Some *Heliconius* males take the act of rape to the ultimate extreme. They probe into the pupa of a different butterfly species and fertilize the female before she's even had the chance to emerge. That always ends as a death sentence for the female.

It's the monarch that most people think of when they have romantic visions of butterflies. The male flies about fluttering his orange-and-black wings in a scene of visual poetry. He spews pheromones, also referred to as "love dust," to attract potential mates. Females eventually float by in a hormonally induced haze, and the male takes his pick, grabbing hold of one of their legs. Unable to fly, they fall from the sky and land on the ground, where the female tries to throw off her suitor. If the male withstands the battle, he lays his head next to hers and strokes it with his knobbed antennas. Once she's been sedated, he flips her over and they mate. He then carries her to the treetops, where they remain until sunrise. The courtship appears to have all the elements of a best-selling romance novel.

"In reality, monarch courtship is brutal and bad," states Gordon Pratt. "They're the thugs of the butterfly world."

Newcomer knew relatively little about butterflies, but he was beginning to understand more about Kojima. He had already guessed there was a dark side to the man.

He had yet to discover that he was in for the ride of his life.

THE U.S. FISH AND WILDLIFE SERVICE

We share this planet with many species. It is our responsibility to protect them, both for their sakes and our own.
—PAMELA A. MATSON

THE 1970S WERE GOLDEN YEARS for the USFWS Office of Law Enforcement (OLE). A growing awareness of the plight of this country's wildlife had grabbed the American public by its shirttails. Federal agents undertook a series of daring undercover exploits in Louisiana to save the American alligator, a species that had been poached nearly to the point of extinction for its skins. Another sting operation targeted the biggest ivory dealer in Alaska, along with a former biologist with Alaska's Department of Fish and Game. This time, the species being plundered was protected walrus. Their headless bodies washed ashore on Siberian and Alaskan beaches, their three-foot-long ivory tusks, and penis bones, having been removed for sale. Included in the arrests was the leader of a drug-dealing motorcycle gang involved in the illegal trade. It was as if the American people suddenly woke up and realized what was happening to the world around them.

Congress passed the Marine Mammal Protection Act in 1972, followed by the Endangered Species Act in 1973. Newly appointed chief Clark Bavin, known as the J. Edgar Hoover of FWS, began to turn old-time game wardens into professional special agents. OLE was in the process of being pulled by its bootstraps out of the Dark Ages and dragged kicking and screaming into modern times. In 1971, 174 game wardens had primarily been banding ducks and checking hunters for unplugged shotguns. That changed in the mid-1970s when wildlife agents found themselves shipped off to Glynco, Georgia, to receive fifteen weeks of intensive training in criminal investigations, firearms, self-defense, and wildlife law. It was their final evolution from duck cops into a new breed of investigators.

By 1977, an all-time high of 220 special agents, trained in the mode of the FBI, successfully broke the back of the illegal alligator trade. The timing couldn't have been better. The exploitation of wildlife was rapidly rising as word traveled of the quick and easy money to be made. More sting operations were undertaken that proved to be successful. Operation Falcon exposed a Middle Eastern plot to smuggle endangered wild falcons from North America for the sport of sheiks and oil-rich falconers. Other cases nabbed ivory traffickers, the illegal sale of eagle feathers, and unlawful importation of sea-turtle products. Operation Renegade broke up a worldwide ring that smuggled scores of rare and exotic birds into the United States. Meanwhile, action was taken at home, as well. Undercover agents busted a group supplying poison to ranchers to kill golden eagles and other critters. Compound 1080, thallium, Temik, strychnine, methomyl, and sodium cyanide were available for sale. Enough poison was found to kill every predator, man, woman, and child west of the Mississippi.

The agents of OLE felt a heady confidence about taking on the challenge as protectors of America's wildlife. They were now federal agents investigating premeditated and well-organized criminal acts that just happened to involve animals. Their numbers were growing and FWS appeared to be solidly behind their work. They couldn't have been more wrong. Though their mission remained the same, it would all be downhill from there.

FWS is primarily known as a biological-research agency responsible for protecting wildlife and its habitats. In a government body mainly composed of managers and biologists, OLE is forever getting a smaller piece of the pie. The number of their agents slowly dwindles, while their investigative caseload continues to grow. These days OLE has 196 special agents. That's down from nearly thirty-three years ago. Take away the number of supervisory people and there are probably only about 130 field agents in total. And fewer than 20 of them do any high-level undercover work. By comparison, the FBI has 12,000 special agents. Yet global wildlife crime ranks just behind drugs and human trafficking in terms of profit.

The FWS budget for fiscal year 2009 was $1.4 billion, out of which the Office of Law Enforcement received a paltry $62.7 million to fight an increasingly sophisticated global war. The illegal trade in butterflies alone nets nearly four times as much as OLE's entire annual budget. It's a sad fact, since special agents are the thin green line fighting to protect endangered species. That line is constantly getting thinner.

"If we turn sideways, you can't see us," Newcomer likes to joke.

In truth, OLE is the most undermanned and underfunded body in the entire federal government.

"We have a lot in common with Rodney Dangerfield. We don't get no respect," lamented former special agent Pete Nylander.

Most people don't even know what FWS agents are or that they even exist. If the public thinks of them at all, it's as game wardens, not as plainclothes wildlife investigators. But then, a good deal of an agent's time is spent buried in paperwork and not out in the field.

"Our laws are very important, or Congress wouldn't have saddled us with them. I think the American people feel the job is being done, and what fools they are," retired FWS agent Terry Grosz sadly declared.

With so few agents and little money, Fish and Wildlife can't enforce half the laws that are on the books. The Office of Law Enforcement is left to work on a triage system. Only the most pressing cases get any attention. Agents tire of the fight and quickly burn out. The best are those who work for the animals and try to sidestep the politics of the agency.

Blowup

No one changes the world who isn't obsessed.
—Billie Jean King

Newcomer went to work Monday morning to find two phone messages that had been left by Kojima over the weekend.

Please call me. We need to talk.

This is Yoshi. We should meet again before I leave.

He tried calling him back without success. He finally dashed off an e-mail.

Sorry I missed your messages. I tried to call this morning but think you might have already left for Costa Rica. If so, please send an e-mail. Otherwise, I'll be near the phone today and tomorrow.

They eventually connected late that afternoon, and Kojima promised to send him butterflies. There was just one catch.

Did Ted remember an auction site called InsectNet.com that

Kojima had mentioned? Under no circumstance was Ted to use Kojima's name, or that of his former company, Butterfly and Insect World, if he ever decided to post butterflies for sale on there. People would know they came from Kojima, and that could prove to be a problem.

Newcomer swore to abide by the rule.

They spoke again later, though a meeting was never arranged.

Kojima flew out of LAX on July 2, bound for Mexico.

NEWCOMER WAS STILL FLYING HIGH at having been taken under Kojima's wing so quickly. He began to work on the next piece of the puzzle.

He placed a call to Immigration and Customs Enforcement (ICE) agent Jamie Holt. Would she mind entering Kojima in their Treasury Enforcement Communications System (TECS)? That way Newcomer would know whenever Kojima entered or left the country. He also asked that Kojima be tagged for random secondary inspections upon flying into the States. What better way to learn if Kojima was transporting contraband in his carry-on bags or luggage? Then Newcomer began to patiently wait both to hear from Kojima and for the parcel of butterflies to arrive.

Kojima finally sent an e-mail three weeks later. He planned to be in L.A. for only one day and then head back to Costa Rica. Would Ted have time to see him?

With bells and whistles on, Newcomer thought merrily. The sooner he got this case rolling, the quicker it could be brought to a conclusion.

However, a flurry of e-mails resulted in confusion as to the date that Kojima would be in town. Kojima's notoriously short patience wore thin. The boy wasn't as professional and on top of things as he had hoped. He didn't have time to waste. Kojima abruptly ended the matter. He was sorry that things wouldn't work out this time around. Maybe a meeting could be arranged for his next trip to L.A., in August.

What was Kojima talking about? Newcomer had absolutely no idea what had gone wrong. The last thing he wanted to do was alienate Kojima. The success of the case hinged on their relationship, but he felt as if he'd just been blown off. And what about the butterflies that Kojima had promised to send him? Newcomer had to become a trusted associate if he was ever to prove butterflies were illegally being brought into the United States. There was no question that a meeting had to be scheduled. Ted Nelson promised to make himself available any time, any place. All Kojima had to do was to name where and when.

But it was as if Kojima had sensed that something was afoot and purposely dodged him.

Newcomer was left to wonder what would happen next. He didn't have to wait long to find out.

He received a general e-mail from Kojima just a few days later. Kojima advised his Internet customers that they should check his Web site, worldinsect.com. He'd brought beetles back from Costa Rica, and they were now up for auction.

Newcomer promptly logged on. Son of a gun! There were the bugs in living color. Posted were probably some of the very butterflies that Kojima was supposed to have sent him. So much for his new best friend. That's when it hit Newcomer like a two-by-four. He was on the verge of losing the smuggler. No way was he going to let Kojima slip out of his hands. He didn't waste a moment but promptly responded.

> *VERY NICE! I love those silver and gold beetles.*
> *By the way, I'll be ready to start selling on e-Bay as soon as you have birdwings and beetles ready to send me.*
> *Have a safe trip to LA and I'll keep my schedule open so that we can meet.*

Let him try and weasel out of that one.

Newcomer didn't hear a word but was left to cool his heels. *Damn the man!* Three weeks later he sent Kojima another e-mail.

He hoped that Kojima's Costa Rican trip was going well and that he was finding lots of interesting insects for them to sell. By the way, the eBay account was up and ready to go. Ted could start anytime Kojima gave the okay. Oh, yeah. There was just one other matter. He needed some butterflies to post.

That ought to do the trick, Newcomer thought as he pressed the button and sent the message into cyberspace.

He was flabbergasted when Kojima responded from Madagascar. Was there anywhere in the world that this guy didn't have access to?

Kojima replied he would return to L.A. soon, but it would only be in transit. As for butterflies, he'd brought some on his last trip to L.A. and had tried to call, but Ted was never around. Too bad he hadn't been available.

It's that damn undercover phone that's screwing me up, Newcomer fumed. Kojima always called him on weekends, or after work hours, when the Fish and Wildlife office was closed.

However, Kojima did leave the door open a bit. He hoped they could do business together soon.

You bet we will, Newcomer promised himself.

He knew what he had to do: apologize whether this was his fault or not.

He was so sorry there had been a miscommunication. It was just that summer was his busiest season, and Kojima hadn't been in town all that long. Ted was ready to start their eBay venture if only Kojima could bring some material.

He ended his note with what he felt was another good excuse for having not been available.

> *I've been seeing a lot of my girlfriend, so I usually stay at her house.*

Kojima was a guy. He would certainly understand that.

Newcomer totally misread him.

Kojima was becoming increasingly annoyed that he couldn't reach Ted by phone. Newcomer was growing equally irritated but for a different reason. He'd arrive at work Monday mornings, check the undercover answering machine, and begin to feel overwhelmed. There were always numerous messages left by Kojima.

What was he supposed to do about the situation? He felt as if his hands were tied. To make matters worse, it was beginning to seem that his personal time wasn't his own. He'd made a vow to Allison and didn't want to create any tension between them. Kojima's calls were an unwanted intrusion. "I'd come in thinking, 'Goddamn it! Why can't I have my weekend to myself?'" he recalled.

At the same time, he realized the problem. Undercover work isn't a nine-to-five job. It requires that agents be on "bad guy" time. One's personal life always comes in second. He'd have to figure out the phone situation and somehow deal with it.

Newcomer began to hear less frequently from Kojima and started to worry that the case was slipping away from him. He needed to know when Kojima next planned to be in Los Angeles. Upping the pressure was that annual training was about to be held in West Virginia. Special agents receive forty hours of mandatory "in-service" law-enforcement training each year, including firearms classes. He would have to attend whether Kojima was in town or not.

He sent Kojima an e-mail, hoping to reignite contact and also divert a potential problem.

Ted Nelson explained that he'd been accepted to the Marine Mechanics Institute and would be in Florida September 8–26 to learn more about boat engines. Was Yoshi possibly coming to L.A. before then?

Newcomer dangled news that he hoped would prove tempting. He'd been checking eBay auctions, and birdwing butterflies were bringing high prices. He was anxious to get started on their site.

Kojima shrewdly waited until Newcomer was out of town to reply.

The boy needed to be taught a lesson. He said he was in L.A. now but had to leave before Ted returned home. Maybe they'd have better luck next time.

Goddamn it! What is going on with this guy? Newcomer stewed.

If Kojima was jerking his chain, he was doing a hell of a good job of it. Kojima was clearly enjoying himself. He knew how to play the game and get what he wanted. It was taking a toll on Newcomer's nerves. The challenge made him all the more determined to bag his prey.

He had a sneaky suspicion that Kojima was lying to him. Newcomer checked with Customs. According to their records, Kojima had flown into L.A. twice in the past few months.

Damn the man. Newcomer suspected that Kojima was no longer bothering to call him. In fact, he wouldn't have been surprised to learn Kojima had made several trips to L.A. that Customs wasn't even aware of. Kojima had claimed to have a U.S. passport under a different name. If that was true, then there'd be no record of when Kojima entered and left the country. At one time he'd had an American wife, a home, and a business in L.A. Knowing Kojima, he'd probably managed to get hold of just such a passport. He'd confided that he frequently switched between the two passports to throw Fish and Wildlife and U.S. Customs off about his travels. There'd be no way to check his comings and goings unless authorities had Kojima's fake American name. Kojima wasn't about to give that out to anyone.

Newcomer had to do something quickly, or this case would never gel. In fact, it would bite the dust. Everything hinged on getting the phone situation resolved. He took his plight to his boss. "This is totally ridiculous. Yoshi can't get hold of me on weekends, and he's beginning to get pissed," he explained to Palladini.

Surely, Fish and Wildlife would listen to reason. Newcomer worked for the U.S. government, not some two-bit, rinky-dink firm. He pressed for an undercover cell phone that would be with him 24/7. But like all federal agencies, Fish and Wildlife is a maddening muddle of bureaucracy. You'd think he'd asked to purchase a stealth missile. It would have to be approved by D.C., and getting anything out of Washington was like trying to pull a rabbit out of a hat. But the FWS Office of Law Enforcement (OLE) was used to that. After all, they'd always been considered the bastard child of the agency.

Newcomer was an eternal optimist, but Kojima was beginning to wear him down. It was now October, and Newcomer was as agitated as a dog with fleas. He still hadn't received any butterflies from Kojima and Fish and Wildlife was hemming and hawing over his request for a cell phone. What was wrong with those people in D.C.?

He had his own spin on an old saying "Desperation is the mother of invention" and decided to take a drive by Kojima's apartment just to see what was up. What do you know? Kojima's truck was clearly visible through the locked gate of the garage.

Was it possible Kojima had been here all along?

Newcomer walked to the front door, tempted to ring his buzzer, but Kojima's name wasn't on the directory. The apartment was listed under one Alexander Denk. Perhaps that was why no one had been able to find Kojima's address earlier.

He went back to work and began to idly troll InsectNet, the Web site that Kojima had mentioned. He did it partly to see what was being sold and partly as a way to deal with his frustration. His Web surfing paid off. Butterflies from Cuba were being offered for sale, and the listing happened to have been posted by Kojima. His Web site was prominently listed as the source. Wasn't that exactly what Kojima had told Ted Nelson not to do? Newcomer felt as if Kojima had been two-timing him. Everything the man had told him so far was a lie. He immediately shot Kojima an e-mail.

He kept it friendly, even though he was secretly annoyed. He'd had enough of the fun and games. Ted Nelson wrote of having seen the butterflies posted on InsectNet. Did Kojima still want to work together on eBay, or had he decided to handle it by himself?

Kojima strung him along with all the finesse of a striptease artist. After all, one never knew. He might actually need Ted Nelson one day.

Kojima was still interested in working together but very busy at the moment. He'd just been to the Paris insect fair, where he'd made eighteen thousand euros in one weekend. He was off to the Frankfurt insect fair next and expected to make a ton of money there, as well. If Nelson remembered correctly, he'd been away the last time that Kojima was in L.A. and had brought butterflies for him. Exactly what did he expect him to do? Kojima figured that ought to put him in his place.

Newcomer read the note and felt like banging his head against a wall. Dealing with Kojima was worse than dealing with a petulant girlfriend. At least a girlfriend would kiss and make up. Newcomer decided it was time to take matters into his own hands. He'd do what D.C. was still dicking around about. It was the only way the case would be saved, and he'd had enough of the bullshit.

He went to a Target store and bought himself a fifty-dollar pay-as-you-go cell phone. He registered the line under the name Donald Duck to make it completely untraceable. The phone could now be used on any of his undercover cases. Even better, he would appear to be even more of a badass if a target ever tried to trace the line. When the name Donald Duck popped, the perp would think Ted Nelson had his shit squared away.

Now I have a damn good reason to reconnect with Yoshi, Newcomer thought. *This time, I'm not even going to press for butterflies.*

Instead, Nelson congratulated Kojima on his success in France and wished him good luck in Germany. He also dropped the news that he had another cell phone number.

> *Remind me to give it to you before you come back to the U.S. It will be easier to get in touch with me when you need a ride from the airport.*

He planned to become Kojima's newest cell phone buddy. Newcomer's scheme seemed to work. Kojima promptly asked for the phone number and said that he'd see him soon.

That was like holding a bone in front of a dog. Newcomer waited a few days, heard nothing more from Kojima, and cracked. He sent another e-mail with a promise to be better about checking messages, and returning phone calls, even from his girlfriend's house. But his undercover cell phone remained silent.

It had now been four months since the two men had originally discussed selling butterflies, and things hadn't progressed any further. Newcomer couldn't figure out what was still wrong. He had bought a cell phone and had fixed the problem.

Newcomer was no fool. He knew the agency's unwritten rule: Screw up and it will haunt you the rest of your career. This was his first case, and it hadn't exactly gotten off to a rip-roaring start.

Palladini assured him that everything was fine. He was just new to undercover work. It would take time for the butterfly case to kick in.

Time was exactly what Newcomer felt he didn't have. At thirty-eight, he had already been a lead prosecutor and lawyer for eleven years. That initially gave him a certain amount of confidence as a special agent. But that boldness was now beginning to decompose.

He was the oldest agent in the Torrance office and found himself stuck in a cramped, windowless room that had previously been used for storage. The view from his open door was a cinder-block wall at the end of the hall. He didn't take it as a good omen.

"The only time I ever saw someone was if they went to the kitchen to make lunch," he recalled, and that wasn't very often.

The room screamed *low man on the totem pole* and felt like the dead end it literally was. It also described how his first case was going. He had to do something to shake things up, or he might very well be a goner.

Perhaps obtaining Kojima's financial records would help provide a clue. If nothing else, at least he'd be taking some action. Newcomer served subpoenas for records on Kojima's two U.S. bank accounts. The Bank of the West had nothing on him. However, Kojima had deposited numerous checks for insect sales in 2002 and early 2003 to his Bank of America account. Since then, there had been no activity.

Newcomer was at a crossroad feeling more and more like a jilted suitor. He'd still received no insects, and Kojima no longer returned his e-mails or calls. All remained quiet on the Japanese front. It only made Newcomer all the more determined.

There's got to be something I can do to get his attention.

Maybe Kojima didn't really believe Ted Nelson could pull off the eBay auctions. If so, he'd have to prove him wrong.

Newcomer trolled InsectNet again while hoping to come up with a plan. There were lots of insects for sale but nothing produced a brainstorm. That is until he stumbled upon a discussion taking place about Kojima in a chat room. Kojima clearly wasn't the most popular guy on the Web site.

An Indonesian dealer complained that he'd sent butterflies to Kojima and never been paid. Other collectors were equally unhappy. They'd tried to buy the butterflies he advertised on InsectNet, but Kojima always claimed that he had sold out. The guy was nothing more than a slick fraud. How else could his prices be so low? They accused Kojima of perpetuating a bad joke at their expense. Or maybe Kojima just hated Americans and that's why he refused to sell to them.

Beware of this man was the phrase repeated over and over.

Newcomer immediately saw a golden opportunity. He slapped on his proverbial "white hat," jumped on his charger, and went riding to Kojima's defense.

> *Whoa! Hold on a moment. I can vouch for my good friend, Yoshi. In fact, anyone interested in obtaining his material should con-*

> *tact me. I've seen it and I'm working with him to make his specimens more readily available to U.S. collectors.*

The word spread like wildfire. Ted Nelson was Kojima's U.S. representative. He'd help supply species that American collectors had never seen before. Just e-mail Ted Nelson and he'd see what he could do for them.

How absolutely brilliant was that? Newcomer thought, thrilled with his plan.

Kojima would be slapping him on the back in gratitude when he realized that Ted Nelson had developed a new customer base for him.

What Newcomer forgot was that Kojima had specifically told him never to use his name on InsectNet.

Newcomer received a flood of positive responses from collectors on the site. They flocked like hungry geese to get Kojima's butterflies and almost everyone wanted Appendix II species. Newcomer wheeled and dealed with all of them. He suggested that one client would get a better bargain if he expressed an interest in buying more than one species, and that another should be willing to pay at least $450 for a pair of *Prepona brooksiana*.

A third collector sent a large list of butterflies that he wanted to get his hands on while someone else asked for a price list of all available lepidoptera Kojima had in stock. Yet another swore he'd forever be in Ted Nelson's debt if he would just get him on Kojima's "want list."

Newcomer laughed to himself. Perhaps he should have been a salesman and had missed his calling. He couldn't believe his luck. He'd hit a virtual gold mine. Collectors would do anything to get their hands on Kojima's butterflies. The guy was a living legend. There seemed no question Kojima would have to be impressed with Ted's initiative and enthusiasm.

Newcomer juggled it all while traveling out of town for Fish and Wildlife, doing his best to keep every ball in the air. He posted a message to collectors on InsectNet. Could everyone just hang on until Ted Nelson returned to L.A. on November 19?

It was then time to claim his reward. Newcomer proudly sent Kojima an e-mail. Not only had Ted set up their eBay auction site, but he'd also cultivated customers interested in buying directly from them. Could they still split the profits fifty-fifty?

Newcomer had no doubt that Kojima would be blown away. Kojima would probably give him the title Director of Business Development and Marketing. He included the list of requested butterflies he'd already received and then sat back and waited to be congratulated.

Newcomer received a reply but not the one he'd anticipated. Kojima had been in L.A. while he was again out of town for Fish and Wildlife. Kojima said that he'd called many times but Ted had never answered the phone. He hoped he had the right number and would try to see him perhaps on his next trip. No mention was made of InsectNet or the list of butterflies that had been sent.

As far as Kojima was concerned, Ted Nelson had once again screwed up.

Newcomer checked his answering machine. Kojima was right. He'd indeed left several messages while he'd been gone. Newcomer carefully listened to each. There was no doubt Kojima was becoming increasingly annoyed and suspicious that Ted Nelson could never be reached.

It was two long weeks before he heard from Kojima again. Newcomer's heart leapt when his undercover cell phone finally rang. Maybe Kojima had a change of heart and decided to take the bait. Once again, it wasn't the response that Newcomer had hoped for. Rather, Ted Nelson proceeded to receive a sound trouncing.

"You have to be very careful about sending e-mails to people you don't know when dealing with CITES butterflies," Kojima warned. "Fish and Wild, and the Department of Agriculture, are always on the lookout for dealers selling without permits. They monitor the Internet in search of amateurs like you that don't know what they're doing."

Kojima grew more agitated as he continued to speak.

Fish and Wildlife would come to Ted's house, check his computer and discover that Kojima had sent CITES butterflies to him by Express Mail. If that happened, Kojima would be the one to get caught.

Then Kojima dropped his bombshell.

He would no longer carry butterflies and insects in his luggage to the States. Kojima had been thoroughly searched by U.S. Customs on his last two trips to L.A. An inspector had gone so far as to specifically question him about butterflies. Why would they do that unless someone had tipped them off? Kojima was now under scrutiny and intended to be all the more careful.

Damn, damn, damn! Newcomer mentally kicked himself in the rear. Sure, he'd asked inspectors to do random checks on Kojima, but he didn't expect them to pat the guy down every time he got off a plane. And why in hell would they purposely ask him about insects? Of course that would make Kojima suspicious and paranoid as hell. Now he knew he was on their radar.

Was there anything else that could possibly backfire on him?

Newcomer took his licking from Kojima and retreated for a few days. Then he continued to push by sending Kojima an e-mail just a week later. Would Yoshi please send some butterflies along with a basic description that could be posted on eBay? Ted specifically wanted CITES material.

Newcomer was overly anxious but felt there wasn't any time to waste. He wanted to get straight to the heart of the matter and going through the motions of selling legitimate bugs would have been a pain in the ass. It meant keeping track of all the money made on unprotected butterflies that were sold on eBay. Why be responsible for a lot of bugs that weren't an issue just to get to the big-ticket items? He wasn't a damn bookkeeper. Newcomer wanted to cut to the chase.

In truth, Newcomer was afraid that he might possibly screw up. The longer the case went on, the greater the risk that he'd lose Kojima. If he did, Kojima's global network would remain in place, and endangered butterflies would continue to be whacked. Meanwhile, all of Newcomer's time and work would have been for naught. He wanted to reach indictable territory as quickly as possible. That would prove to be a fatal mistake on his part.

Kojima's reaction was to respond with a "Dear John" e-mail. Kojima had decided against doing eBay but wished Ted Nelson good luck with his own business.

No! Newcomer nearly screamed aloud as he read the note. Kojima couldn't pull out of their venture now. He had to do something to quickly salvage the case.

Newcomer wrote back and did his best to apply a good dose of guilt. What about all those people he'd found that wanted to buy Yoshi's butterflies? He couldn't let them down now. Kojima's reputation was at stake. Besides, he'd gone so far as to check out the new potential clients on an Internet detective site. None were agriculture or wildlife agents. No way would they make trouble. Couldn't Yoshi just send him the requested list of CITES butterflies? He'd like the opportunity to try and sell them. Ted Nelson would accept all of the risk. Newcomer threw in another approach while he was at it.

> *By the way, I fully agree with your decision that eBay is not the way to go. I'd much rather develop a private list of customers to sell your material to. I've been looking forward to this for a long time. Why end now?*

But Kojima refused to be swayed. He didn't trust Ted's new "clients." The butterflies they wanted were expensive and difficult to obtain. Someone would realize that Ted knew nothing about butterflies and report him to Fish and Wildlife. Then agents would come asking questions about import permits and where he got his material. Exactly how would Ted Nelson answer them? Kojima would be left a sitting duck.

The more Kojima wrote, the more he stoked his own fire.

Ted Nelson knew nothing about the business. That was precisely why Yoshi had called so often and tried to meet with him in L.A. But Ted was always away or with his girlfriend. Meanwhile, Kojima was the one with everything to lose.

Newcomer tried to be reasonable but the quarrel continued to escalate. He argued that he would have gladly sold butterflies on eBay if Kojima had only provided material.

If Ted Nelson wanted a fight, Kojima was more than happy to oblige. The bickering was akin to two boxers duking it out. Kojima now promptly delivered a jab to the head, a right cross, and an upper cut to the chin.

Ted only wanted to make money. He had no interest in the nuts and bolts of the business. Did he know anything about pricing and how to handle rare butterflies? How did he plan to get payment from his customers? And what would he do if a butterfly broke in shipment and that was the only one in stock? None of those issues had ever been discussed. He was already taking orders without any material on hand. Ted Nelson had absolutely no experience in the business of selling butterflies.

Newcomer made one last attempt. He'd read of butterfly collectors that had gotten into trouble because they'd sold to undercover agents. Ted would never sell to anyone that he hadn't checked out first. In addition, he would never tell who supplied him with butterflies.

That proved too much for Kojima who flew into a rage. Wasn't that exactly what Ted had already done on InsectNet? Ted Nelson had openly declared himself to be Kojima's U.S. agent.

He now went in for the knockout. Ted was greedy and not willing to put in the work, he was too busy having fun with girlfriends, and Kojima was tired of running after him.

Kojima finished by completely spelling it out for Ted. He would no longer keep an apartment in L.A. or sell butterflies in the States because it had now become too dangerous. Their business relationship was over. It wasn't going to work out.

Kojima cut him off cold. There was no resuscitating the patient. Ed Newcomer had run his first undercover case into the ground.

Butterfly Madness

"Mad" is a term we use to describe a man who is obsessed with one idea and nothing else.
—Ugo Betti

Newcomer was devastated. He was unused to failure, and losing Kojima came as a major blow.

He broke the news to Marie Palladini. He had screwed up big time. She advised him to relax and take a deep breath.

"I told him to just wait it out. Kojima would come back. I knew the history of the guy. We don't like anybody to get one up on us. We're such a small agency that our players will sometimes resurface if we miss them," she recalled.

Newcomer was grateful that Palladini handled so well what he considered to be a fiasco. Even so, he continued to be vexed by the case. He replayed conversations in his mind and wondered what he could have done differently. One thing kept leaping to the forefront.

Maybe if I hadn't acted so quickly. If I'd just asked for meaningless, common butterflies, maybe I would have caught Kojima by now.

"I had fantasized about being an agent for so long. This job felt so right that I never had any doubt I could do it," Newcomer acknowledged.

He'd been ultraconfident. Now he had to learn to move on. He threw himself into another case and tried to put Kojima behind him.

Newcomer hoped to prove that Aeroflot crew members were sneaking beluga caviar into the country in a consistent and calculated fashion. It was late 2003, and at that time there was an exception in the CITES rules. People were allowed to bring two small tins of beluga into the United States inside their personal luggage.

Newcomer did a detailed analysis of declaration forms and noticed an odd pattern among crew members on Aeroflot flights. Either the entire seventeen-member crew brought in the allotted amount of caviar all at the same time, or none of them did.

Newcomer suspected that smuggling was afoot. Most likely the crew checked in to a local hotel, combined their tins of caviar, and sold the entire lot to a local dealer. Forty thousand dollars' worth of beluga caviar was brought into the country during a six-month period in this way.

The case looked great on paper, however the U.S. Attorney's Office turned it down. There was one major problem. The CITES loophole screwed things up.

"All someone had to say was, *I brought the caviar in for my own personal use and never ate it. So, what else were we going to do with all this caviar? Throw it away? Of course, I didn't intend to sell it at the time of import,*" Newcomer explained.

He was back to where he had started.

There were plenty of cases to keep Newcomer busy, but Kojima continued to haunt him like a phantom pain. He kept hoping to receive an e-mail or a phone call, only to wait in vain. Everywhere he looked there seemed to be butterflies flying in all directions—except toward him.

PROBABLY NO ANIMAL OR INSECT has come to represent the process of transformation more than the butterfly. Shrouded in mystery for centuries, they're considered to be fragile symbols of metamorphosis, life and death, beauty and hope. After all, what other species has the chance to be born twice? And how many of us wish we had that opportunity? Except for the ugly duckling, there is no more powerful example of metamorphosis.

Butterflies have fascinated man from the beginning of recorded time, carrying with them heavy spiritual and emotional significance. They've been found on Egyptian tomb frescoes and on both Roman coins and funerary monuments. A butterfly emerging from a human skull was discovered in a mosaic in the ruins of Pompeii; while the Greeks use the same word, *psyche,* to describe both butterflies and the human soul. Not only were butterflies a symbol of the spirit in early Christianity, but the idea of reincarnation was conceived as Brahma watched caterpillars turn into butterflies. The bugs can be found in every country and culture.

Chuang-tzu, a Taoist poet, philosopher, and mystic, was so taken with butterflies that he once dreamt that he was one. He awoke uncertain if it had been a dream or if he was actually a butterfly now dreaming of being a man.

Native American legend portends that if you whisper a secret wish to a butterfly, it will be flown to the heavens and your wish granted. Before a race, Navajos rub the colored dust from a butterfly's wings on their legs in order to run swiftly and light with its spirit.

Mexicans believe returning monarch butterflies are the souls of dead children, while people in the Middle Ages imagined that butterflies were really fairies.

Children imprisoned in Nazi concentration camps used their fingernails and pebbles to carve hundreds of butterflies into their barrack walls. It was as if they knew they were going to die and hoped their souls would become butterflies. According to eyewitness accounts, butterflies fluttered around Auschwitz and other death camps for years after the Holocaust. Many believed the children's souls had been freed at last.

And in China, single white butterflies have been found in the cells of recently executed convicts. All had converted to Buddhism as one of their last requests.

The *Papilio indra* has a devoted following verging on the maniacal in America and Japan and was the focus of the first successful butterfly sting operation in California. As fate would have it, the case agent had been John Mendoza, the man that Kojima outwitted. Newcomer knew the case inside and out. He'd read all about it during his preparation for the L.A. Insect Fair.

Fish and Wildlife had received a tip back in the nineties about Richard Skalski, a pest exterminator at Stanford University nicknamed "the Bugman." He was known to be poaching protected butterflies in national parks, wildlife refuges, and national forests and selling them in the commercial trade. Included in the lot were rare *Papilio indra panamintensis* and *Papilio indra kaibabensis* from Death Valley Monument and Grand Canyon National Park.

Working with him was Tom Kral, a young real estate assessor. While Skalski was described as an emotionless neat freak, Kral was filled with enough emotion to fuel a jet engine. Completing the group was fellow poacher and associate Mark Grinnell. At fifty years old, Grinnell still lived with his parents in a room that should have been disinfected, scrubbed, and cleaned as part of his punishment.

The trio spent a decade doing the job of three environmental wrecking balls. They lived, breathed, and dreamed of nothing but butterflies, documenting their passion for the winged insects in a series of four hundred letters, many of which were signed "Yours in Mass Murder," "Yours in Crime," and "Yours in Poaching." Naturally, they kept their correspondence, which was eventually found by Fish and Wildlife.

Skalski's collection of Kaibab swallowtails was reputed to be the largest of its kind in the world. He was absolutely gaga over the little black butterflies. Patches of blue on their hind wings danced before his eyes. The dabs of color resembled sprinklings of fairy dust, each punctuated by a single fiery red flame.

The trio was also obsessed with precisely labeling their prey. Every tiny butterfly sported four toe tags containing the pertinent information: where the bug was caught, by whom, and exactly when. It made the prosecutor's job of convicting the men that much easier.

Like most trophy seekers, the men relished the thrill of the hunt. Skalski would obsessively drive from Redwood City, California, to the Grand Canyon in one night, fueled by six-packs of Coca-Cola, along with his passion for *indra*. Meanwhile, Kral took twenty thousand envelopes on one collecting trip, fully expecting to fill them all up. He was disappointed to return after having gathered only nine thousand Rocky Mountain butterflies.

The adrenaline rush came not just from the hunt but from then taking their quarry home with them. All the while they reveled in knowing the prey was theirs as they spread and pinned each butterfly's wings. Sterile, glittering equipment routinely lay spread on Kral's dining room table as though he was a surgeon. The numerous pieces were specifically used for mounting his catch. It was part of the ritual.

"It's beyond a passion. It becomes fanatical. Obsessed collectors have to have every single type, every single specimen, and cost is no object. They want something that's easy to dominate. It's a control issue. They have the power of life and death over butterflies," explained Fish and Wildlife entomologist Chris Nagano.

Skalski and Kral went to great lengths to keep their collecting spots hidden, revealing them to only a select few. "I'm entrusting you with my secret! Do not share it with anyone," Skalski once wrote to Kral.

"Because some of the things you sent me are on the endangered species list, I will be careful not to reveal where I got them . . . it's best to trade 'under the table' like this," Kral responded.

"It's like controlling DeBeers diamonds," Nagano explained. "They kept their locations secret so that they were the only ones to have access to certain butterflies."

Kral pretended to be a bird-watcher, or simply claimed ignorance of the law if caught while collecting. He'd quickly stash away his portable butterfly net, known as a "National Park Special."

Fish and Wildlife agents found one of the finest collections of butterflies ever seen during their raid on Tom Kral's home. They spent eleven hours combing through what Nagano estimated was more than 100,000 butterflies, of which 1,637 species were illegal. But a far eerier discovery awaited Special Agent Mendoza at Richard Skalski's house.

He entered Skalski's kitchen and opened the refrigerator door to find something he'd never seen before. Mendoza froze at the sight of a newly emerged butterfly, still alive but entombed in a glassine envelope. The light and heat in the room worked its magic as the butterfly started to awake from its slumber. Its feet began to move as though it understood its fate and was struggling to escape. The volume mysteriously grew until it erupted into a Greek chorus and the crackling of glassine envelopes filled Mendoza's veins like ice water.

His horror intensified as he slowly realized that hundreds of newly emerged butterflies were neatly stacked inside, each held captive in its own pristine cell. All were alive, though they'd been imprisoned for weeks. The cold, dark refrigerator kept them in a semi-dormant state until the internal fat in their wings metabolized. Then they'd be thrown into Skalski's freezer to die. He'd be left with perfect specimens that hadn't flapped off a single scale.

The butterflies now reacted en masse to the sun and warmth by stomping their "feet," the only appendages they could propel. Thousands of little legs joined in the struggle to be free of the glassine envelopes. The sound was the stuff of nightmares, as if thousands of nails frantically scratched within prematurely closed coffins.

An even more bizarre sight awaited the agents in Skalski's bedroom, where dozens of *indra* pupae were taped to the ceiling above his bed; that way he could lie awake and watch, excited as a child to unwrap a new gift. The little mummies were the last thing he saw at night and the first thing he saw when he woke in the morning. He eagerly waited for each butterfly to emerge. They would know only a moment's freedom before being confined to death within transparent glassine walls. Approximately eighty-seven *Papilio indra kaibabensis* swallowtails were found in Skalski's home. It was the largest collection of its kind in the world.

Catching and possessing perfect butterflies was what Skalski, Kral, and Grinnell lived for. Altogether, 2,375 butterflies were confiscated from the three men, including 14 of the 20 North American butterflies that were listed under the Endangered Species Act at that time.

"They're hunting trophies, but we're hunting them. They're our trophies," commented former agent Pete Nylander. "In that sense, we're more like them than many people realize."

Skalski received five months in jail and a fine of three thousand dollars. Kral and Grinnell were given each three years probation, a three-thousand-dollar fine, and ordered to perform community service. But other collectors have been known to be equally illegal and greedy.

"There's an unspoken code of ethics that you don't really want to collect more than you need. Every now and then you get an individual who not only takes hundreds of specimens but also cuts down all of their food plants so that other collectors won't be able to get them. This is where you cross the line into the pathological," explains entomologist Ken Osborne.

"Some of these people you find are pirates," butterfly dealer Ken Denton agrees. "They're outlaws and they're collecting for commercial gain. Their incentive is to try and make all the money they can. They don't care about the laws and the rules."

Denton believes the focus should be more on observing butterflies and less on collecting them, as mounting habitat loss occurs. "People look at butterflies as a replenishable resource, but that's not actually the case."

Is it possible, then, to separate a pure love of butterflies from the obsessive urge to collect them? That seems to be the challenge. It was a fascination with butterflies that captivated collectors in the first place. Perhaps the lesson to be learned is that even butterflies should not be possessed.

The eBay Scheme

Collecting, apparently, is a lifelong obsession.
—Miyeko Murase

Newcomer had become haunted by butterflies and, no matter how hard he tried, couldn't give up the ghost. Kojima had worked his way under Newcomer's skin to reside in his very bones. Newcomer was determined to get Yoshi back somehow. There was no way he'd let his first undercover case become a dud.

He was drawn once more to InsectNet, hoping to learn whatever he could about Kojima's butterfly sales and activity. He soon found what he was searching for.

It was February 2004, and Kojima was once again offering to sell CITES Appendix II butterflies on InsectNet's "classified" section. With Kojima in Japan, and out of touch, there wasn't a damn thing Newcomer could do except continue to monitor the site.

Kojima posted forty-six different species of butterflies for sale, collected in Asia, the Caribbean, and South America, over a five-week period. He was raping the globe to provide the best and rarest butterflies for his customers. Just the thought of him getting away with it continued to eat at Newcomer.

It was serendipity when his confidential informant called that April from out of the blue. He had recently spoken with Kojima by phone.

It was true that Kojima no longer brought butterflies into the States, but the CI had a new gem to offer. Kojima had bragged that friends in California were now supplying him with *Papilio indra* butterflies.

The informant told Newcomer that he was still willing to assist in Fish and Wildlife's investigation. He would gladly make undercover purchases from Kojima.

Newcomer was newly invigorated about the case. Perhaps this was the best way to catch Kojima; then again, perhaps not. Kojima was not only clever, but he had an ego the size of the Amazon. That might very well prove the key to his undoing.

Newcomer was used to scrambling back up after having been knocked down so much as a kid. Get beaten up enough, and you learn to dodge, punt, and bluff. The trick was remembering to go with the flow. In a way, Kojima was a lot like those bullies that he'd once had to deal with. Newcomer felt ready to jump back into the game. All he had to do was pinpoint Kojima's Achilles' heel.

Newcomer now had the brainstorm that he'd been waiting for all along. He'd pretend to find another butterfly supplier and run his own eBay auctions. That would prove to Kojima that he could sell Appendix II butterflies without being detected by Fish and Wildlife.

The idea was brilliant, and Newcomer was sure he knew just how Kojima would react. His plan would not only work like a charm; it would drive Kojima insane. He would be so jealous that he'd come running back and they'd be in business together again. Besides, at this point, Newcomer had nothing to lose.

He ran the scheme by Palladini and got it approved. Then he contacted eBay. Newcomer explained the situation, told them he wanted to hold decoy auctions and needed to create a fake history. Would eBay help out? They flatly refused.

"I thought, 'Fuck them,' and did it anyway," Newcomer said with a laugh.

The first thing he needed were protected butterflies to sell. That was the easy part. He obtained Appendix II butterflies that had been previously seized by Fish and Wildlife. The next challenge was to ensure that none of them were really sold to the public. That's where his fellow agents from around the country came in. Newcomer contacted some Fish and Wildlife buddies and had them sign up for decoy eBay accounts. It was then time to put his plan into action.

Rigged auctions were set up that ran each week based on an elaborate system. Newcomer contacted participating agents and had them place incremental bids on whatever butterfly he had posted. That would drive the price up an exorbitant amount, creating what appeared to be a real bidding war. One agent would enter an absurdly high bid at the end of the week and be named the winner.

He offered tantalizing tropical specimens, from *Ornithoptera paradisea,* often called the "most desirable birdwing the world over," to the delicate *Ornithoptera meridionalis*, to the dusky black *Troides aeacus* with its sun-kissed hind wings. All of them were protected. Every agent that "bought" a butterfly would post a glowing review.

Wow! Ted's butterflies are fantastic. This guy's the greatest!

The goal was to attract Kojima's attention. Each night Newcomer would fall asleep not counting sheep but imagining what Kojima might say.

I see you're doing eBay auctions, Ted. Where are you getting your material? Who's your supplier?

While dozing off, Newcomer would pretend to smile and shake his head.

Yoshi, you know I can't tell you that. It's all hush hush. But he's not that good. He's very unpredictable, and charges too much.

Don't worry. I can supply you, Kojima would always whisper.

Newcomer would have sweet dreams all night long.

The eBay auctions were such a success that Ted Nelson appeared to be making a killing. Newcomer used the disc of butterfly photos that Kojima had given him just to add fuel to the fire.

"The beauty of it was that members of the public were also bidding. Of course, none of them ever won," Newcomer said, with one exception.

An agent screwed up and didn't enter a high-enough bid one week. A member of the public won an *Ornithoptera victoriae* from the Solomon Islands. The butterfly is as beautiful and exotic as it sounds. Large and colorful, its wings are a pulsating lime green boldly trimmed in jet black. Three golden spots dot each outer edge. Newcomer had to think fast and cover his ass.

Fortunately, the winner lived in Southern California, and Newcomer offered to deliver the butterfly in person. Then he quickly set to work. He painstakingly slit the butterfly's wings with a knife. To his mind, it was better to ruin the specimen than sell it and help fuel demand. He presented the butterfly to the man along with an apology.

"I totally screwed up. I was folding the wings and one tore. I'm afraid you're not going to be able to use it," he explained ruefully.

Newcomer offered to return the man's money and asked only that he not post negative feedback. The customer agreed, and Newcomer was back in business.

His auctions gathered steam, but they had yet to attract Kojima.

For chrissake, what do I have to do to get this guy's attention?

Then he hit upon it. If the mountain wouldn't come to him, then he'd simply go to the mountain. Kojima still posted bugs for sale on InsectNet, so Newcomer proceeded to write up an announcement.

TED NELSON NOW SELLING BUTTERFLIES ON EBAY

I'm pleased to announce my association with a reputable but-

terfly collector. Introductory auctions will start at ridiculously low bids. I know you'll like the product.

He posted it prominently on InsectNet, along with his link. He figured that ought to do the trick.

He figured right.

Boom! Kojima went ballistic.

Ted Nelson immediately began to receive e-mails from Kojima. However, they weren't congratulatory notes. *"Stop using my photos immediately!"* he demanded. *"Otherwise appropriate action will be taken."*

Newcomer got a good chuckle out of that one. *Gotcha,* he thought. He further fanned the flames by responding that Kojima had been right all along. The eBay business was huge, and he was doing very well. Perhaps Kojima would like to meet with him the next time he came to L.A. and discuss buying and selling butterflies.

That ought to give him a good tweak. Newcomer made the concession not to use Kojima's pictures anymore. Even so, Ted Nelson's eBay auctions continued in full swing, and he let Kojima know that he planned to run many more. Wouldn't Yoshi reconsider their business relationship?

What else could Kojima possibly want? Not only had Newcomer proven that he could successfully sell butterflies, but he was inviting Kojima to participate.

"I want to work with you," Newcomer wrote.

But he had once again misread the man. His ruse backfired as Kojima became even more hostile. He now viewed Nelson as a competitor attempting to take his territory. And he began to retaliate.

Every time Nelson auctioned a butterfly, Kojima would offer the same exact species on InsectNet.com, only at a lower price. Kojima undercut him on everything from *Ornithoptera goliath* to *Morpho cypris* to *Bhutanitis mansfieldi.* It was a war of the wings. Included in his postings were public scoldings of Nelson.

Shame on you, Ted Nelson! You're selling CITES material without permits!

I sell cheaper than Ted. Don't buy from Ted Nelson! Ted Nelson is bad!

The turn of events would have been hysterical if it wasn't so maddening.

The guy is freaking nuts. He's like a pit bull on steroids, Newcomer complained to himself. Kojima was accusing Nelson of doing exactly what he always did, selling CITES Appendix II species without permits. It became a game of chess in which every move Newcomer made was countered by one from Kojima. Newcomer knew he was walking a fine line. How far could he push Kojima and still stay within the boundaries of the law?

"You're always trying to manipulate someone's behavior through your own in such a way that you can collect evidence. At the same time, you have to be very careful not to entrap that person," Newcomer explained.

The battle heated up as Kojima kept a watchful eye on Ted Nelson's eBay auctions while Newcomer monitored Kojima's sales on InsectNet. The men circled each other like two angry bulls snorting and puffing. Newcomer's plan had brought Kojima out of the woodwork, but it wasn't getting him anywhere. There had to be another way to lure the man.

It was then that Newcomer realized what had been in front of him all along. Kojima was so angry that he'd begun to get careless. He was now offering Appendix II butterflies to every Tom, Dick, and Harry on InsectNet. He was breaking his own Golden Rule: Only deal with people you know, and you'll never get caught.

Newcomer now knew his next move. He'd let Kojima step into his own trap.

Ed posted a *Bhutanitis lidderdali*, a species from Bhutan and southwest China, on his eBay site. The butterfly is a showstopper, black bedecked with narrow ribbons of cream color. However, its real beauty lies in its hind wings, which are fringed with a series of pointed tails. The colors are the last gasps of a sunset, three white orbs, reminiscent of crescent moons, separated by bands of red and gold.

Kojima not only followed suit, he upped the ante by offering three of the same Appendix II butterfly on InsectNet as "This Week's Special."

It was too good to be true. *Kojima has actually helped set himself up,* Newcomer gloated to himself. Kojima was practically doing his work for him.

Newcomer promptly contacted San Diego agent John Brooks with a request. Would Brooks mind buying all three butterflies on InsectNet from Kojima?

Brooks couldn't have been more thrilled. It seemed they were finally going to catch the bastard.

Brooks sent an e-mail to Kojima through InsectNet, and the deal was done. There was just the matter of payment. Kojima preferred to receive the $137 due via PayPal. Newcomer nixed the idea.

"Let's find out how Kojima cashes his checks. Tell him that a money order works best, and ask where you can send it," Newcomer instructed.

Brooks followed through.

Kojima responded that he still had family in California and provided an address in Beverly Hills.

"He is my father-in-law. Don't worry. He likes money," Kojima jokingly added in his e-mail.

That worked for Newcomer. What could be better than catching two birds with one stone? The father-in-law apparently had plenty of time on his hands and was helping Kojima with his business in this way.

His father-in-law cashed the money order, and the butterflies arrived a few days later by international Express Mail from Japan. Inside were three *Bhutanitis lidderdali* packed in glassine envelopes, their wings neatly folded as if in prayer.

Also included was a U.S. Customs declaration form on which Kojima had marked the butterflies as a gift with zero value. With it was a fraudulent CITES permit that not only had expired three years ago but was issued from the wrong country. Kojima was unbelievable, yet he'd gotten away with this sort of thing for years.

Newcomer now had Kojima on a number of charges. Kojima was guilty of smuggling, a false Customs declaration, false labeling, failure to declare, and violating the Endangered Species Act. Throw in the fact that his father-in-law might be considered a coconspirator, since he'd accepted the money order and cashed it. There was just one problem.

Kojima was probably the world's greatest butterfly smuggler. No way was Newcomer about to go to the U.S. Attorney's Office and indict him on a $137 butterfly-smuggling charge. It would be a waste of both the U.S. Attorney and Fish and Wildlife's valuable time for what amounted to nothing more than a minimal offense. Newcomer wanted Kojima thrown into jail and put out of business, not merely slapped on the wrist.

He continued his selling on eBay if for no other reason than to piss off Kojima. By the end of the summer, he had run fifty-two auctions. Best of all, eBay never knew an agent had manipulated its system after it had nixed the opportunity to cooperate.

John Brooks attempted to contact Kojima once again in the hope of making a few more buys, but Kojima no longer responded. His attention was apparently focused elsewhere. For a while Newcomer wondered if Kojima had simply disappeared, until he received one last angry e-mail.

Kojima accused Newcomer of still using his photos and threatened that "big trouble" would soon be coming. Ted Nelson wouldn't be in business much longer. He signed the e-mail "Not your friend, Yoshi."

It proved to be a timely warning.

The California Department of Fish and Game (CDFG) received an anonymous call on its "Cal Tip" hot line at the end of June 2004. A caller reported that Ted Nelson was engaged in the sale of illegal butterflies in Los Angeles.

The wardens contacted the U.S. Fish and Wildlife office in Torrance to have them investigate. As luck would have it, Ed Newcomer answered the line.

"They told me, 'Hey, we got this funky recording from a Japanese guy. He's reporting that someone named Ted is illegally dealing in butterflies," Newcomer remembered.

It was as if the floor had suddenly fallen out from under him.

He asked them to make a copy of the tape and send it to him right away. He anxiously waited for the package to arrive and then listened to the recording in stunned silence. There could be no doubt about it. He immediately recognized the voice. Kojima had turned him in to California Fish and Game.

Newcomer's first response was to laugh out loud in surprise. Son of a gun. Who would have dreamt that he'd make such a gutsy move? Little did Kojima know that he'd just turned in a federal agent.

Newcomer's second reaction was giddiness. *This is totally awesome. My cover is 100 percent tight.* The bad guy believed it enough to try and turn him in.

Then a knot the size of the Rock of Gibraltar formed in Newcomer's stomach as reality began to set in. *Oh shit. The case is blown,* he realized.

Newcomer ran into Palladini's office and relayed what had just taken place.

"I screwed it up again," he confessed. "I'm never going to get Yoshi."

Palladini agreed this time. "You're right. It's over. Kojima turned you in. He's never going to trust you after this. He'll figure you've either become an informant or made a deal with Fish and Game if you stay in the butterfly trade. I doubt you'll ever hear from him again."

Kojima had adroitly marked his territory and just gotten rid of his competition. Newcomer had been bumped from the game. Turning Ted Nelson in had been a slick move. Newcomer had been thoroughly beaten. He'd driven the butterfly case headfirst into the ground for the second time.

Yoshi Kojima was the cleverest and most paranoid man that Newcomer had ever known.

Operation High Roller

You've got to get obsessed and stay obsessed.
—John Irving

The year was 1973. Hundreds of species had already vanished from the planet. Extinction caused by man had been ongoing for centuries, from the dodo to the Tasmanian tiger, the Barbary lion, the great auk, and the Carolina parakeet. Human greed and consumption decimated the Steller's sea cow, a creature the size of a large truck weighing in at three and a half tons and measuring thirty-five feet in length. Hunted relentlessly, the species became extinct only thirty years after having been discovered, exterminated for its meat and skin.

Passenger pigeons flew across the eastern United States in numbers so large they darkened the sky, only to be wiped out by "market hunters." John James Audubon claimed to have seen a flock in Kentucky a mile wide that took three days to pass through. He calculated there were three hundred million flying overhead per hour. The storm of their flapping wings could be heard as far as six miles away. But even they went the way of the dodo. The last of their kind died on September 1, 1914, at the Cincinnati Zoological Garden.

More species were rapidly disappearing when the United States passed what many hailed as the most powerful conservation law in the world. The Endangered Species Act (ESA) would help put the brakes on extinction by requiring federal wildlife agencies to protect threatened and endangered species while preserving their habitats.

"It was two hundred years ago that there were buffalo on the plains, there were whales in the ocean, and there was an attitude that we can do no wrong," former prosecutor Lee Altschuler told the judge during the Skalski, Kral, and Grinnell butterfly case. "And it was twenty-two years ago that the Congress came to a different determination."

They had decided to pass the Endangered Species Act.

On paper, the ESA sounds like one hell of a big club for the agents of Fish and Wildlife to wield. In truth, it has all the clout of a wet noodle.

Its biggest flaw is that ESA criminal provisions don't include felonies involving serious jail time. Rather, the most severe crime under the act is a Class A misdemeanor, which means someone can kill the last tiger in the world and the worst that will happen is a yearlong jail sentence. Low criminal penalties are one reason the illegal wildlife trade is booming.

Less known, but with far more teeth, is a statute called the Lacey Act. The Lacey Act was drafted by Iowa congressman John Lacey and introduced in the House of Representatives in the spring of 1900. The act's original purpose was to ban the interstate trade of illegally hunted game and waterfowl. The law has since been amended a number of times to include not only protected wildlife but also insects, plants, and timber. The Lacey Act now prohibits the importation, exportation, sale, and purchase of wildlife and plants in violation of state, federal, Indian tribal, or foreign law. The last condition is the very best part. The United States has the authority to enforce another country's conservation laws whether that country chooses to uphold them or not.

The maximum penalty under the felony portion of the statute is $250,000 and/or twenty years imprisonment. It's helped expand the role of federal wildlife agents and is now one of the broadest and most comprehensive forces in their arsenal. The Lacey Act has become an important weapon for combating international wildlife crime. Due to this, the statute has its detractors, who claim that the law is misguided and see it more as a catchall akin to Big Brother or the Patriot Act.

A few wildlife dealers complain that the Lacey Act has given FWS everything it needs to go after any and everyone, and that agents have turned into thugs. Many believe U.S. citizens shouldn't be forced to follow another country's laws when that same country doesn't appear to care about them. They grumble that the Lacey Act makes it all the more difficult for businesspeople to import foreign butterflies and bugs and that the law is confusing.

Not so, dissents Brent Karner, associate manager of entomological exhibits at the Natural History Museum of Los Angeles County. He believes the Lacey Act to be perfectly clear in intent, as well as an absolute necessity. "The laws are in place to protect the natural world so that there will be something left on this planet for people to enjoy and to learn about," Karner declares.

Kojima knew the law well, but he never let that stop him. He was the man who could never be caught.

Newcomer didn't hear another word from Kojima after his last threatening e-mail on June 17, 2004. Neither did John Brooks or Newcomer's confidential informant. It was as if he had dropped off the face of the earth. He was a spider that had gone back inside his hidey-hole. Kojima didn't even show up at the 2004 L.A. Bug Fair. Newcomer felt like a total failure. Kojima had managed to outfox him. Kojima's network was bigger and better than ever, and he was still out there somewhere selling butterflies. That was the biggest irritation of all.

"Mendoza missed him and then I missed him. Kojima's left thinking he's the most wanted butterfly smuggler in the world and that he'd outsmarted the Fish and Wildlife Service. That totally pissed me off," Newcomer recalled.

He had no choice but to become immersed in other work as two years rolled by. Newcomer investigated sea otter shootings, trafficking of illegal piranha, and Jackson chameleons illicitly flowing into the mainland from Hawaii. He eventually dipped his toe back into undercover work while on a case involving Asian arowana smuggling.

Asian arowana, or dragon fish, are native to Indonesia and Malaysia and considered to be the rarest and most prized freshwater fish in the world. Beautiful and mysterious, they can sell for as much as ten thousand dollars apiece on the black market. Red ones are believed to ward off evil spirits from one's home, while gold arowanas will bring their owner good fortune and wealth.

That wasn't the case for the smugglers Newcomer dealt with when he posed as a middleman with connections to a corrupt U.S. Customs official. He played the go-to guy who could get anything into the States. His target was a crooked Indonesian supplier. Newcomer's "job" was to receive the endangered arowana and make sure that they passed through Customs. He'd then deliver them to illegal buyers. It wasn't long before the pipeline was swiftly closed down, though the final outcome wasn't the one that Newcomer had hoped for. The case was indefinitely shelved at the U.S. Attorney's Office while the Environmental Crimes Unit went through some shuffling and reorganization. In the meantime, wildlife crime continued its onward march.

Even so, the arowana case helped bolster Newcomer's confidence after his disastrous blowup with Kojima. He now figured undercover work was the way to go.

"I learned I could do it and play other people. I wasn't nervous or afraid now that I realized I could pull it off," Newcomer said.

His ongoing work also convinced him that wildlife criminals don't love animals in any way, shape, or form, no matter what they might say. They're in the business of buying and selling animals for only one reason. Every wildlife crook he'd ever known had tried to convince him otherwise, but Newcomer concluded it was all a pile of crap. They loved butterflies, fish, birds, or whatever else they sold, as one would love diamonds for their value.

Special Agent Lisa Nichols succinctly summed it up. "It's just to make money. The rarer, the more unusual, the fewer, and the weirder something is, the more it's in demand by people in the trade." That was why Kojima remained the king of butterfly smugglers. He continued to obtain what no one else could.

Newcomer tried to put the case behind him and not constantly dwell on Kojima. It drove him crazy whenever he did. So much so, that Newcomer couldn't bring himself to close the case. He convinced himself it was low-priority and not worth the time it would take to fill out the necessary paperwork.

In truth, an obsession is hard to shake, and his defeat continued to eat away at him. Kojima had become Newcomer's Moby-Dick. The folder lay like a slow-burning ember buried beneath other cases that sat on his desk. The only concession Newcomer made was to slap a Post-it note marked CLOSED on its cover.

He finally moved into a larger office in 2006 and was able to make it his own. Piles of papers and folders mushroomed on his desk like weeds growing out of control. They jockeyed for space with books touting the titles *Hand-to-Hand Combat*, *Basic Stick Fighting*, and *Special Weapons/Special Tactics Training*.

By March, Newcomer found himself sitting at his desk with a bag of Fritos and a jumbo cup of McDonald's Coca-Cola while wondering what to do next. He needed to tackle something totally different from the butterfly case and stop feeling like such a failure. At the same time, he craved the addictive rush that only a big case could bring.

Family photos stared down from his bookshelf, seemingly unaware they were being stalked by a plastic model of Bigfoot. Newcomer glanced at his old boyhood Smokey the Bear thermos, still without a scratch or a mark, when his eyes came to rest on a plastic bag of eagle feathers. He munched on his breakfast of champions as a memory began to stir.

He'd received a call back in the fall of 2003, just as the Kojima case had kicked into gear. A man had found a wounded red-tailed hawk in his backyard and taken it to the California Wildlife Center. A staffer proceeded to notify Newcomer. The hawk had been injured by gunshot. It was the kind of case that agents usually don't take, since they tend to be unsolvable.

Even so, Newcomer called the man who initially found the bird. Did he happen to know anything about it? His response was no, but Newcomer's timing couldn't have been better. He'd just discovered a second hawk in his yard that very day. This one was dead. In fact, he'd thrown the bird into his freezer.

Either all hell had broken loose, and birds were haphazardly falling from the sky, or something wasn't quite kosher in North Hills. Newcomer's interest immediately perked up. He arranged to pay the man a visit.

He drove out near the Van Nuys airport to the scene of the crime. The area was a sprawl of tract houses all lined up in rows. Looking at the neighborhood, he figured the perps were probably a bunch of kids idly plinking at birds.

Newcomer knocked on the man's door and asked if any teenagers lived in the neighborhood. Then he began to point at individual houses and inquire who lived in them. Also, what did each of the residents do? He finally pointed to the house next door and asked the same questions.

"Oh, that's Marty Ladin," the man said. "He loves birds. In fact, he breeds pigeons."

Bingo! If Ladin loved pigeons, then it was a good bet that he hated hawks. Newcomer instantly had a suspect on his list. He didn't immediately go and talk to the guy. First he interviewed Ladin's neighbors.

One woman told of hearing gunshots one day and finding a large bird with big yellow feet in her yard. Another neighbor had a similar experience. Comparable tales were repeated throughout the neighborhood. Gunshot blasts always resulted in dead birds being found in someone's yard. By now, a siren was blaring inside Newcomer's head. It was time to head over and interview Marty.

The first thing Ladin wanted to know was what all the neighbors had said. Something else struck Newcomer as odd. Marty was a little too eager to cooperate in any way possible. If he'd been any more helpful, Newcomer would have had to pin a deputy's badge on him.

He brought the conversation around to pigeons, and Ladin soon began to babble like an overflowing fountain. He revealed that he had racing pigeons worth five thousand apiece and hawks were killing his birds. By the way, what would happen to someone who snuffed out a hawk? he asked.

Newcomer listened as he studied Ladin's setup. A bunch of shotguns lay all around the grounds. Ladin followed Newcomer's gaze.

"You carry a weapon?" he asked Newcomer.

Ladin's sinister tone was enough to creep Newcomer out. He knew Marty was the guilty party, but without hard proof there was nothing he could do about it at the moment. That's where the survival skills he'd honed as a boy came in handy.

He paid Marty Ladin another visit a few days later. This time he wasn't empty-handed; he came armed with two videotapes. He laid them on the kitchen table like a couple of six-shooters. Then he asked Ladin a few more questions.

Had Marty ever noticed the streetlamp that stood outside his house?

Ladin answered yes, though he wasn't quite sure what the hell that had to do with shooting a hawk. Newcomer quickly filled him in.

"There's a camera inside it. We've got you on tape shooting that bird," Newcomer informed him.

Ladin folded like a bad hand of cards at a poker game. He admitted his crime, pled guilty in court, and paid a five-thousand-dollar fine.

Oddly enough, neither he nor his lawyer ever asked to view the evidence. It was a good thing, since Newcomer had managed to pull off one of his "brass ball" bluffs. There was no camera in the streetlamp, and both videotapes had been blank.

Newcomer now remembered something else Marty Ladin had once told him. Other guys in the neighborhood belonged to pigeon-racing clubs. In fact, there were pigeon clubs all over Southern California. That could mean only one thing. Hundreds of pigeon aficionados were probably having the exact same problem with hawks. More than likely, they were trying to get rid of them, too. Newcomer quickly did the math. If that was true, then thousands of hawks and falcons were being killed each year along one of their major migration routes, the Pacific Flyway.

Newcomer had grown up in Colorado as a child of the seventies. To see a hawk during that time had been rare. Spotting a raptor was akin to glimpsing a grizzly bear in Yellowstone. If you were lucky enough to spot one, you'd stop whatever you were doing and stare. Newcomer spent his childhood thinking that hawks were cool. He also knew something about their history. The use of DDT had affected the numbers of falcons and hawks in the United States from the 1950s to the 1970s, causing their populations to plummet. It had taken time, money, and years for the raptors to make a recovery. The thought that people would wantonly harm them now totally pissed him off.

He knew the Ladin case had been only the tip of the iceberg. The subculture of pigeon racers was a whole other world. It had remained on a back burner in his mind while the Kojima case had been active. Newcomer suddenly knew the next case that he wanted to do, but he'd first have to sell it to Marie Palladini.

He did his homework in order to present a rock-solid case. His first discovery was that investigating racer-pigeon clubs while working undercover would be a problem. These clubs were hard to infiltrate; there was very little social interaction, and members usually didn't go to one another's houses. Instead, the birds were released at different locations and their flight times noted upon return.

However, Newcomer learned of another, equally popular sport, one that featured Birmingham roller pigeons. Referred to as "spinners" and "tumblers," the birds are bred with a genetic defect that causes them to roll in midair. The pigeons fly to great heights, where they experience a brief seizure, throw up their wings, flip backward, and repeatedly somersault, looking like feathered whirling dervishes as they spiral uncontrollably toward the ground.

The good ones live to compete another day. The unfortunate ones end up meeting their maker.

Roller-pigeon fanciers fly their flocks in "kits" composed of twenty or more pigeons. Seeing them tumble in syncopation is akin to viewing performing dolphins at SeaWorld or watching a new Olympic team sport. For hawks, they're the raptor version of Kentucky Fried Chicken on the go. Roller pigeons are viewed as fast food on the wing.

Newcomer remembered having first seen roller pigeons in the Denzel Washington movie *Training Day*. Gang members flew them as a warning whenever cops entered their neighborhood. He'd also heard that the boxer Mike Tyson was an avid roller-pigeon devotee.

Newcomer put his notes together and presented the case to Palladini. She was less than enthusiastic at first. Special agents have plenty of endangered animals to save. Hawks were way down on Fish and Wildlife's list of priorities."I just didn't see it as a federal case," Palladini explained. "I thought there were a lot bigger cases we could be making with Newcomer's talent and ability. Why would we waste our time on this when the state could be doing it?"

As it turns out, the state could never have done what Ed Newcomer did.

He pleaded his case by arguing that the number of hawks being killed impacted the resource and made it a high priority. Palladini reluctantly gave the okay but warned him not to spend too much time on it. She first wanted to see where the case would lead.

Next he had to break the news to his wife.

He'd first met Allison at a tae kwon do tournament back in Washington State. He was working as the assistant attorney general in Olympia, and Allison was doing social work in Seattle. He'd immediately shared his goal of wanting to become a U.S. Fish and Wildlife special agent. Even so, Newcomer said that he'd never do undercover work. They dated for six years and eventually married.

Newcomer had gone back on his word with the very first case. However, dealing with Kojima had never truly felt like an undercover operation. Infiltrating the roller-pigeon crowd would be a different matter. Many club members were African Americans from tough East L.A. neighborhoods. Newcomer was more Rocky Mountain High. He slowly eased Allison into the idea by talking about it as the case got under way. She was carefully prepped as to the change that would take effect. Even so, this wasn't what she had signed up for.

"I remember thinking I hadn't really understood that he'd be going undercover, and here he was getting ready to enter an inner-city group wearing a gun and a wire," she said, tucking a renegade curl into place. Her long wavy hair was carefully contained in a ponytail. "I knew he'd be carrying a gun, but it was still very different from what I had expected. I married a lawyer and ended up with this guy that packs an AR-15 rifle and a shotgun."

Things would work out as long as he continued to honor their pact about family time. Otherwise, what was the sense of even being married?

Still, it was weird to be the wife of a guy who put his life at risk. At the same time, she knew Newcomer was doing a job that he totally loved; the work was important, and he was good at it.

There was another thing that struck her as odd about Newcomer getting into criminal work. Upon moving to L.A. they'd taken an acting class together for fun. Newcomer always joked about being such a terrible actor and not knowing how people could do it. Yet here he was not only taking on different personas but also being able to fool everyone.

"I thought it was weird that he was such a good liar. It was like, 'Hey, wait a minute,'" Allison admitted. It was a quality that she didn't know he had, and she wasn't quite sure what to make of it.

Newcomer had a simple explanation. "I learned to be a good liar as a kid in order to defend myself against shitheads."

Newcomer morphed into Ted Nelson once again, having had plenty of time to perfect his undercover persona. Nelson had a Social Security number, credit cards, and even an undercover bank account, all newly acquired since the Kojima case.

He was now ready to make contact. Newcomer searched the Internet for information on pigeon clubs in the area and phoned one of the listed members. The fancier invited him to attend an Inner City Roller Club (ICRC) pigeon show to be held within the next few weekends. The show's location would be in a lot behind the Pigeon Connection, a bird store at Western and Eighty-third, on the edge of South Central Los Angeles.

Newcomer would have to change his appearance. You couldn't go looking too white bread when you're invited to a pigeon show in South Central L.A., and you certainly couldn't appear to be a clean-cut, buttoned-down cop. Most roller-pigeon competitors are blue-collar guys, so Newcomer grew a long droopy mustache and threw on a ratty ball cap to adopt a gruffer look. He also exchanged his crisp clothes for scruffy boots, beat-up T-shirts, and a pair of jeans that he found at a Goodwill store. The new Ted Nelson didn't have his own commercial marine business; instead, he made deliveries for one. Newcomer wired himself up with a video camera and headed to the show.

A large, racially mixed crowd milled about the lot. One man

flaunted a Pomona gang tattoo on the back of his head, while another brandished a Rollin' 30s Crips tattoo on his neck. The men normally wouldn't be caught dead or alive in one another's neighborhoods. Yet here they raced pigeons, drank beer, and talked together. Newcomer was warmly welcomed as a newbie who was eager to learn about birds.

"Part of it was that I was the new mark, the guy you could maybe sell your shitty pigeons to," Newcomer explained.

Everyone spoke openly about killing hawks. A man bragged of having recently whacked a peregrine falcon with a pump-action pellet gun. There were lots of ways in which the birds could be killed. It was literally choose your own poison.

One vendor sold a poison paste that he'd concocted. A little Temik mixed with mayonnaise and rubbed on the back of a pigeon's head worked wonders. Put it on one or two sacrificial birds that weren't flying very well, and simply wait for a hawk or falcon to take a nosh. Presto! The attacking bird was a goner.

Another guy sold pigeon jerseys for ten dollars a pop. They were little vests that had Velcro on the front. The backs were constructed with loops and loops of fishing line. All you had to do was to put one on your pigeon and let a falcon strike. Once its talons became entangled in the snares, the bird would panic and drop to the ground.

"At that point, you can take all your frustration out on the falcon," the vendor explained.

He emphasized his point by pretending to brutally stomp on a bird. Of course, this was only his part-time job. The vendor's real career was working as a smoke jumper for the U.S. Forest Service. He put his life on the line to save the forest and then literally stomped on its inhabitants. The irony took Newcomer's breath away.

Nelson was also advised on the best way to dispose of a dead hawk's carcass. The smart thing was to bag it up and throw it into a Dumpster several miles from your house. That way you wouldn't get caught. After all, killing hawks and falcons was illegal and could incur a ten-thousand-dollar fine. But not to worry; it was nearly impossible to be discovered trapping and killing them. He was then shown all the hawk traps that could be bought. They were large contraptions made of wood and wire mesh that measured about twelve feet square.

Newcomer spent six hours bonding with his new best friends. One of them took his photo and posted it on ICRC's Web site the following day. Ted Nelson was listed as a "newcomer" in the caption. Newcomer found that to be hilarious. But he also knew that they'd have to be caught in the act before he could arrest anyone. That required Marie Palladini's approval.

He showed her the surveillance tape. Palladini listened as the men professed their hatred for falcons and hawks and bragged of killing them. She also saw the number of traps that were for sale. He told her that the president of the national group, the National Birmingham Roller Club, lived right here in L.A., along with a local chapter of 250 members. If just 50 percent of them killed 10 hawks each year, then 1,250 hawks annually were being slain in this area alone. That amounted to a huge impact along the Pacific Flyway.

The next club fly was taking place at a number of members' homes, including that of national president Juan Navarro. As far as Newcomer knew, they were all killing falcons and hawks.

That proved enough to convince Palladini. The problem was much worse than she had originally thought. She gave him the go-ahead, and Newcomer went at it full steam. He was determined to target those in leadership positions as well as the worst offenders, most of whom would be at the next fly on April 15.

Ted Nelson now became part of the roving band that traveled to members' homes to watch their roller pigeons being flown. It was clear the men all had a bond. The setting was as social as a tea party.

Newcomer took note of the wood-and-wire traps that were set near everyone's pigeon lofts. Each trap contained a separate bottom cage, where a couple of live pigeons had been placed as bait. They were all designed and sold by Darik McGhee for the bargain price of $120 apiece. He built them in his garage. McGhee explained that he liked to give something back to the hobby and support the people involved. Besides, the more people that bought traps, the fewer hawks there would be.

A security guard at a roller rink, McGhee took it upon himself to become Ted Nelson's mentor. Extremely aggressive pit bulls were kept chained in his yard; the female had started to dry up and wasn't producing enough puppies, and McGhee casually mentioned that he'd soon be "getting rid" of her.

He also told Ted that he'd recently shot a hawk with a .177-caliber pellet gun equipped with a scope. McGhee proceeded to pick up the gun and show how it was done, shooting one of his own pigeons by way of demonstration. It was no big deal. The bird hadn't been flying very well lately, anyway.

McGhee used to have a five-gallon bucket filled with the trophy talons of all the hawks that he'd killed. He'd recently disposed of it, having realized that they constituted evidence.

Another member of the group wore a black T-shirt with a drawing of a peregrine falcon on the back. The caption beneath it read: "Wanted Dead or Alive (Preferably Dead)."

Peregrine falcons had been listed as endangered in 1975, with just 324 nesting pairs in North America. Federal and state agencies as well as conservation groups had mounted a huge campaign to boost their numbers. The falcons were finally removed from the Endangered Species list in 1999. Now here was this bozo flaunting a T-shirt advocating killing them. Just looking at it made Newcomer feel sick.

Everyone was taking photos of the goings-on at each person's house, and Newcomer snapped a few of his own. He managed to catch the license plate of the man in the peregrine falcon T-shirt and ran a criminal history on him. What do you know? It just so happened that he was wanted for rape in Los Angeles County. The quandary was, how could Newcomer turn the guy in and not blow his cover?

Newcomer placed an anonymous call to the LAPD and reported when and where this guy could be found the next morning. Sure enough, the L.A. County Sheriff's Department showed up and arrested him.

The news consumed the members of the pigeon fly at Juan Navarro's house the following weekend. They'd all known about the rape charge but couldn't figure out how their friend had been caught. It was then that Newcomer realized hawk killing was considered more than a badge of honor; it constituted a brotherhood that demanded a code of silence. You didn't turn in a member just because he was wanted for rape, robbery, or drug running. Besides, a few of the men had their own criminal histories.

That didn't include club president Juan Navarro, who lived in a three-million-dollar mansion in a swanky neighborhood near Griffith Park. Located in the hills above Hollywood, it's one of the largest urban wilderness areas in the nation, home to not only the Hollywood sign but also mule deer, mountain lions, coyotes, peregrine falcons, and Cooper's and red-tailed hawks.

Navarro openly bragged of killing approximately forty hawks each year. He was unable to shoot them, since his neighbors were so close. Instead, he'd slide a stick in the trap and ping the hawk on its head. Once it was dazed, he'd quietly pummel it to death.

"It's a great thing. You'll get all your frustrations out. You'll see," he told Newcomer.

Only one person seemed suspicious of Newcomer: Brian McCormick, president of the Riverside County California Performance Roller Club.

The next fly took place at his home the following weekend. The place was easy enough to find; McCormick had a large pirate flag flying from his flagpole. Newcomer strolled the grounds and noticed a baited hawk trap set under a giant tree in his yard. He proceeded to ask him about it.

"I'm just starting to get into this, but I understand I'm going to need a hawk trap. Does it work better under a tree for some reason?" he inquired.

It was a bad move on Newcomer's part.

McCormick glared at him and tersely responded, "I don't trap hawks. What are you talking about? Anybody tells you they trap hawks is stupid."

Newcomer realized his mistake and quickly backed off. "You don't? Okay, I got it. No big deal."

But he knew McCormick was up to no good as the man brusquely walked away.

Newcomer then spotted a Remington 870 shotgun custom-fitted with an eight-foot barrel. It leaned against the house near where McCormick stood with a group of men. Newcomer sidled up in time to hear a remark made by McCormick. Noise from the nearby highway made it difficult for neighbors to hear gunshots that came from his property. Rather, they sounded like trucks backfiring, while the long barrel helped to act as a silencer.

McCormick glanced around, and his eyes momentarily locked on Newcomer as he added that the "po-po" might be watching him. Newcomer understood that to mean the police and now realized that he was under suspicion.

McGhee called a few days later and warned Ted to be careful when it came to talking about hawks. "You're brand new. Some of the guys get a little touchy when you start asking questions about them," he advised.

Newcomer rushed to his own defense. "Whoa! That's all any of you talk about. I didn't even know I had to worry about hawks until I started getting into this hobby. Now I find out that I have to buy a hawk trap or they're going to kill all my birds. You want me to learn about raising pigeons, don't you? Well, isn't this part of it?"

McGhee smoothed his ruffled feathers. "Don't worry, man. I told them you're a good guy. I know you. I'm your mentor."

It was great to know his cover was solid, but Newcomer couldn't help but be upset. The men liked to pretend they protected their pigeons the same as they would any pet.

"That's what really got me," Newcomer admitted. "It was total BS. Every single one of these guys broke a bird's neck when it got old or didn't perform to their satisfaction. They'd throw it in the garbage or over a wall and wait for raccoons and cats to take off with them."

Newcomer ended every pigeon fly by driving to the nearest McDonald's, turning off his surveillance equipment, ordering a jumbo Coke, and making notes of what he'd observed. Before long, he'd amassed an enormous cache of material but had yet to snag anyone.

It was while teaching defensive tactics to FWS agents in Washington that he bumped into fellow special agent Dirk Hoy. The two began to compare notes. Hoy told of a case he'd just opened in Oregon involving "a really weird group of people that fly these pigeons." He was pretty sure they were killing falcons.

"Oh, my God. I'm doing exactly the same thing," Newcomer told him.

They decided to coordinate their efforts and work together.

There was no question that roller-pigeon fanciers were hammering hawks flying across the western United States. The question was, what was happening to raptors throughout the rest of the country?

Newcomer sent an e-mail to every FWS agent in order to find out. In it, he explained what he was doing, what he'd discovered, and that there were clubs all over the United States. Any agent interested in starting a case should contact their resident agent in charge. Newcomer also asked that they coordinate with him and Hoy before pursuing any action. All cases would have to be taken down on the very same day so that no evidence was destroyed.

Newcomer had hoped for positive feedback. Instead, his initiative promptly backfired.

"Ed approached agents from other districts because he had this vision of where the case was going; but he didn't go through the proper chain of command, and for that, he was reprimanded," explained Palladini.

One of the regional FWS supervisors took it as a personal affront and responded by saying, *"Who the hell does Newcomer think he is, telling my agents what to do?"*

Once again politics in the U.S. Fish and Wildlife Service trumped expediency. Newcomer began to have trouble on another front, as well.

Bill Carter, then chief of the Department of Justice's Environmental Crimes Unit, disagreed with Newcomer's strategy. He wanted the first man on whom there was solid evidence arrested and the case brought to a quick end. It would look bad if FWS allowed the killing of hawks to continue just to make a splashy case. The government could be perceived as creating additional crimes by sanctioning such a move.

Newcomer stood his ground, seeing this as their chance to hit hard and send a law-enforcement message for generations to come.

"If only one guy is arrested, then everyone else in the clubs will think, 'Wow. Too bad he got caught. Boy, was he stupid. I'm sure glad it wasn't me.' The culprit will receive a slap on the wrist, and no one's behavior will change. We need to show that killing hawks is condoned by the clubs," he reasoned.

Newcomer was adamant about taking down as many offenders as possible. It was a controversial move, but this was their only shot. The fact was, an undercover agent wouldn't be able to work his way in with these groups again for years after this. At the same time, he heard the clock ticking away. He had a limited amount of time before he'd have to close the case down.

Carter gave him a short reprieve, and Newcomer ratcheted up his undercover work.

He'd recently heard of Rayvon Hall, a man who'd bought one of McGhee's hawk traps and was a member of Brian McCormick's pigeon club. Ted Nelson decided to pay him a visit. He stopped by and introduced himself as a friend of McGhee's. He'd heard Rayvon had great pigeon lofts. Would he mind if Ted took a look at them?

Hall cordially agreed to show him around.

They went to the backyard, where Hall, like every other roller-pigeon fanatic, had a hawk trap set. Nelson casually asked how he disposed of the raptors. Hall explained there was a school nearby so that shooting the birds wasn't usually an option. The sound of gunshots could result in complaints to the police. Therefore, he'd devised his own favorite method of dispatching them.

"I mix bleach with ammonia, put it in a spray bottle, and spray it in their face and mouth. The chlorine gas burns their eyes, along with the inside of their lungs and throat." He liked to watch as they slowly flapped around, suffocated from the poison, and died.

Newcomer was momentarily stunned. It was one thing to quickly kill an animal, but this was torture, pure and simple. Rayvon Hall had just put himself on Newcomer's short list. This guy was going down.

It was one reason why Newcomer watched the TV show *Animal Precinct,* even though it bothered him intensely. He felt it was important to know what people were capable of. It also fired Newcomer up to do his job. Hall's actions were right up there with some of the worst episodes.

"Then I cut off their talons and throw the bird out," Hall added.

"Oh, yeah? Wow! I'd really like to see one of those talons," Newcomer said, giving the man all the rope he needed to hang himself.

Hall had recently killed a Cooper's hawk, and the talon was somewhere around. He found it and handed it to Newcomer. "I give them to my friends as souvenirs. Here, you can have it," Hall offered.

"Really?" Newcomer asked.

This was one of those gotcha moments. He'd bought a hawk trap from McGhee and gotten him to admit that it was illegal. Now Hall was giving him a talon. The evidence was beginning to mount. A few more hits and he'd have a home run.

Next on his list was Keith London, owner of the Pigeon Connection and president of the Inner City Roller Club.

Newcomer parked his Chevy Tahoe across the street from London's residence early one morning. He was there to conduct surveillance with a fellow agent. Timing is everything in life. Today it worked in Newcomer's favor. London not only had a hawk trap on the roof of his garage, but you could see the damn thing from the street. Even more amazing was that a hawk had just been caught inside and was frantically flapping its wings.

The two men sat, waited, and watched. London soon appeared and climbed the roof with a pump-action pellet gun in his hand. He pumped the pellet gun and shot the bird once, but that wasn't enough. London pumped it a few more times, carefully took aim, and shot the hawk again.

Newcomer felt as disconnected as though he was silently watching a movie. Part of him wanted to jump out, stop the killing, and help the bird fly off. He could take the case down right then, though little would have been accomplished. London would receive a one-count misdemeanor, and the case would be blown without having had a deterrent effect. Thousands of hawks would have died for nothing.

Instead, the gotcha moment set in as Newcomer realized, *I've got to get these guys. London is going to pay.* He couldn't believe that he was actually getting the killing on videotape. It was the only way he could deal with the event playing out before him. A flash of excitement rushed through him as he registered that London was finally nailed.

The agents remained glued in their seats as London reset the trap and tossed the dead hawk into his backyard. Then he climbed down, grabbed the bird, and disappeared inside the house. He emerged a short time later carrying something wrapped in a white paper package. London angrily tossed it into his trash can, got into his vehicle, and drove off.

Newcomer waited until London was out of sight and then scrambled to retrieve the booty. Inside the package was a freshly killed Cooper's hawk, its carcass bloodied with apparent trauma to its head and chest. The bird was limp and still warm to the touch.

It was one of those episodes that Newcomer truly hated. He'd made a personal pledge never to kill an animal, not even while working undercover. Newcomer had gone so far as to become vegetarian. Yet he'd sat as a hawk's life was taken and done nothing to stop it. Doing so would have brought the entire undercover operation to a screeching halt. Any action on his part would have not only jeopardized cases in California, Washington, and Oregon but prevented a number of dirtbags from being caught.

He took a deep breath and held his emotions in check. The hawk was bagged and tagged as evidence. The carcass would be sent to the USFWS Forensics Laboratory for a necropsy.

Newcomer now determined that surveillance and night work were the best way to go. That meant he spent his days at the office and nights doing spot checks on pigeon breeders from 10 PM until one in the morning. Then there were the weekend pigeon flys to attend. Spend that much time on the job, and it definitely puts a strain on your marriage. Newcomer was finding that out firsthand. Tension began to build as things became rocky on the home front. Still, Newcomer had no choice but to keep up his hectic schedule.

He asked Special Agent John Brooks to help him out once more. Next on his list of nighttime stakeouts was Brian McCormick's place. He hoped to find proof that McCormick was baiting hawk traps.

Newcomer and Brooks donned camouflage gear, grabbed a couple of AR-15 rifles and two pairs of night-vision goggles, and arrived in an unmarked vehicle. They made their way onto an empty field adjacent to McCormick's backyard.

The night was peaceful and calm as they crept toward the fence. Then all hell broke loose. McCormick's dogs came tearing outside snarling and barking. Two seconds later, the lights flicked on, illuminating the pirate flag flying against the night sky in his yard. The skull and crossbones couldn't have appeared more sinister. Newcomer and Brooks took five giant leaps and dove into the tall grass for cover as a figure stepped out of the house.

It was a warm night, and Newcomer lay perfectly still looking at the stars. The sky was filled with them. There were more than he'd ever seen before in his life. So many that he nearly became mesmerized. Then he remembered McCormick's Remington 870 shotgun with its extended barrel and his nighttime reverie quickly dissolved. At the same time, he realized something else. The two agents were lying motionless in rattlesnake habitat. The area was riddled with the reptiles.

He continued to listen for the sound of McCormick's footsteps in the grass, but it was hard to hear anything above the barking of the dogs and the pounding of his own heart. Even the rush of his blood beat like a kettledrum in his ears.

The dogs finally stopped barking, McCormick went back inside, and the lights were turned off. The agents decided not to press their luck. Brooks and Newcomer packed it in for the night and vamoosed with their backsides still in one piece.

The next target that Newcomer wanted to focus on was national president Juan Navarro's house. Newcomer called in the big guns for help on this one.

Sam Jojola, a twenty-five-year FWS veteran, was an undercover agent extraordinaire. His personal mantra defines exactly how he operates: *It's all smoke and mirrors.*

Jojola had been a member of an ultra-elite group of deep-undercover investigators, part of Fish and Wildlife's Special Operations unit. The squad had consisted of five special agents who conducted long-term international investigations. The strike team focused on those dealers illegally exploiting the most fragile wildlife resources in the commercial trade.

Jojola worked everything from bird-smuggling and reptile cases to big-game poachers. He'd had at least ten different covers over the course of his career and always managed to keep them separate. Jojola joined the military when he was eighteen years old, became a paratrooper, and was assigned to an Army Ranger unit in Ft. Benning, GA. After that, he worked as a New Mexico state prison guard before turning to wildlife. Jojola had experience with people of every ilk.

"It's a chess game, and the pieces are the people," Jojola explained. "In this game, you're moving four, five, and six steps ahead; you have to lay the groundwork for developing the right moves early on. Make the wrong move, and everything shuts down. Then you're in checkmate."

Jojola once morphed into Simon Calderon for an undercover sting on illegal hunting guides. Calderon's personal vehicle was a beat-up van complete with a shot-out windshield. He made sure to register the vehicle in his undercover name. It was something not all Fish and Wildlife agents think to do. That proved to be a smart move on Jojola's part.

An illegal-hunting guide had a friendly police contact run a search on Calderon's license. If Jojola had been a fed, it would automatically return as "no record found." Instead the license came back registered to one Simon Calderon. Jojola went so far as to have medication filled in his undercover name in case the guide looked inside his shaving kit.

Jojola's favorite cover was Nelson DeLuca, owner of Silverstate Exotics, dealing in rare reptiles and birds. Silverstate Exotics was Jojola's front for a sting called Operation Chameleon.

The suspect in the case accused Jojola of being a Fish and Wildlife agent whenever he saw him. He even held a poisonous Gila monster up next to Sam's face as a threat. Jojola made his chess move by reaching up and gently scratching the reptile's head.

"That's the amazing thing. Sam continued with the case no matter what. I don't know if I would have stayed with it once it became that scary. The guy's either playing or testing you, so how you react is key," Newcomer said.

Jojola knew the key well and exactly how to work it. You had to believe the character you were playing, and be someone that your suspect would want to spend time with.

Jojola was one of the living legends of Special Ops. He spoke fluent Spanish, flipped money, bought birds, and ran real businesses as part of his sting operations. He was a true master of the undercover trade. These days he was working as the deputy resident agent in charge of Fish and Wildlife's Torrance office. He was also a strong advocate of Ed Newcomer.

"Ed has everything going for him," Jojola stated. "He has the technological, legal, and undercover capabilities, along with strong organizational skills. Some people may have two out of three, but Newcomer has them all plus the political smarts."

Newcomer asked if Sam would help with surveillance on Navarro's place. Jojola instantly agreed.

The two men planned their course of action, jumped into an undercover vehicle, and drove to Juan Navarro's late one evening.

When you're driving around Los Angeles in the dead of night, everything looks like a giant movie set. A huge *Tyrannosaurus rex* appeared to have torn its way through the roof of Ripley's Believe It or Not Museum. The former Max Factor Building, the Magic Castle, and Hollywood High were all ready and waiting for their next screen take. The Hollywood United Methodist Church stood bedecked with a giant pink ribbon along with a sign that read, ALL ARE PRECIOUS IN GOD'S SIGHT. Griffith Park loomed nearby.

Navarro lived in a large Tudor house on a swanky street in Los

Feliz. The two agents had just turned a corner when a private security guard appeared smack on their tail. Flashing his yellow lights, he pulled them over. Dressed all in black with knapsacks, .40-caliber guns, and night-vision gear, Jojola and Newcomer could easily have passed for a couple of professional prowlers. Newcomer explained the situation, and the security guard went on his way.

The agents continued around the block before parking on a quiet corner. Setting off, they hiked in a few hundred yards toward a culvert. Then they crawled on their hands and knees in a no-man's land. Tall blades of dry grass brushed against their cheeks in a blizzard of bristles, and each weed crushed beneath them resonated as loud as a scream.

They finally hit a drainage ditch that served as a border between two residential neighborhoods. The concrete canal was about fifteen feet wide and filled with assorted debris and muck that turned their feet into suction cups. Newcomer and Jojola slogged through the waste until they arrived outside Navarro's backyard. There they pushed aside the overgrown foliage to find the perfect view. Newcomer quickly spotted a baited hawk trap ready and waiting for its next victim, and the men went to work.

Surveillance cameras were set in the ditch and aimed precisely through the chain-link fence at the trap. Triggered by motion, the cameras would record whatever activity took place. If Navarro was caught in the act, then a point would have been made: The president of the National Birmingham Roller Club condoned the killing of falcons and hawks.

All the while, the men heard rustling near their feet . Jojola flicked on his night-vision goggles and spied giant rats scurrying along the fence line. The agents finished setting up their equipment and left on the double. Then they drove once more to the front of Navarro's house.

Two black garbage cans had been placed at the edge of his drive since their arrival. Garbage cans also appeared in front of the other houses on the block. Apparently, tomorrow was trash-collection day.

Jojola looked at Newcomer and made a suggestion: "Going through trash is a great investigative tool not used nearly enough by agents. I say we go for it."

Newcomer agreed, and they parked a short distance from the house. Scurrying back, they grabbed the garbage cans, threw them into the pickup, and drove off.

"We got them and Navarro didn't see us!" they said gleefully, giving each other a high five.

They discussed where to take their booty as Newcomer drove down the street and swung onto a busy boulevard. The conversation was abruptly interrupted by what sounded like the bouncing of three giant basketballs.

Ba boom, ba boom, ba boom!

Newcomer glanced in the rearview mirror in time to see the last garbage can flying out of the truck.

"What was that?" Jojola asked in alarm.

Newcomer's head fell onto his hands. "You're not going to believe this, Sam, but all three garbage cans just fell off."

"I guess we should probably go back and get them," Jojola replied dryly.

By now, not only had the lids popped off, but the contents of the plastic bags inside were strewn all over the road. It made no difference to the cars that dodged and wove around the containers without slowing down.

Newcomer did his best to stop traffic as they scooped up what garbage they could. From there they drove to a nearby lot, where they spread the trash on the ground in their own variation of Dumpster diving. People milled about, but no one paid them the least mind. They were just two more guys hanging around in the middle of the night.

Jojola and Newcomer spent the next few hours digging through the refuse of the Navarro household. By the time they were through the men knew exactly what the family had eaten that week. However, no evidence of hawks or falcons was found.

The garbage cans were closed back up and dropped off in front of Navarro's house. They'd try another garbage run in a couple of weeks. It would take that long to get the stench of refuse out of their clothes and nose. The agents wouldn't have been surprised if a flock of gulls had followed them home.

RESURRECTION

Whenever anything is being accomplished, it is being done,
I have learned, by a monomaniac with a mission.
—PETER DRUCKER

NEWCOMER WAS FLYING HIGH. It had been six weeks since he'd begun Operation High Roller, and things were starting to heat up. He was in tight with the pigeon people, and every weekend was spent at a club fly. There was still a list of targets that he wanted to bring down as hawks continued to be killed. He was in a delicate balancing act that could land him in Fish and Wildlife political hell and endanger his career. At the same time he knew it was the right thing to do and wouldn't let the threat of that stop him.

He was in his office plotting his next move when the telephone rang. He answered and immediately recognized the voice of his confidential informant from the Kojima case. His nerves began to thrum right after the first word had been said.

"You're not going to believe this, but there's a rumor that Kojima's coming to the 2006 L.A. Bug Fair," the CI told him.

The news hit like a bolt of lightning from out of the blue. Newcomer had thought that he'd finally put the case behind him. Even so, there was no denying a familiar twinge of excitement, though he also felt somewhat confused. Kojima had said in no uncertain terms that he'd no longer attend the L.A. fair. He'd become tired of nitpicking with small-time dealers. Something must have changed.

"By the way, he's not going to have a booth this year," the CI continued. "From what I hear, Kojima's bringing specific butterflies to sell to some of his special clients."

To Newcomer's mind, that meant the butterflies had to be both expensive and illegal. It was the only thing that would make the trip from Kyoto to Los Angeles worthwhile.

"You're sure he's coming?" Newcomer asked. His fingers were already twitching to find the Kojima file. It was buried beneath a mountain of folders on his desk.

"Yep. As far as I know he'll be there," the CI confirmed, and hung up.

Newcomer impatiently pushed aside a mound of paperwork that cascaded in domino fashion onto the floor. He barely noticed as his hand grasped the file. He pulled it out and stared at the yellow Post-it note, with its message CLOSED burning into the cover. He'd changed in so many ways since his last encounter with Kojima. That had been two long years ago, and he'd been a naïve young agent at the time. He was now a seasoned pro who loved undercover work. There were times when he felt more comfortable playing a role than he did in his own skin.

Newcomer slowly peeled off the Post-it note and crumpled it in his fist, as every butterfly now seemed to take up residence in his stomach. Maybe, just maybe, he was being given another chance to redeem himself. It was worth taking a shot to find out. Besides, he was curious to see Kojima.

The fair was on the weekend of May 20 and 21, less than five days away. He'd have to think of some excuse to give McGhee and the others, as to why he'd miss the upcoming pigeon fly. Ted Nelson, wholesale marine-business owner and butterfly entrepreneur, was about to be resurrected.

Newcomer was determined to come face-to-face again with Yoshi

Kojima, the man who had turned him in to California Fish and Game as a smuggler.

NEWCOMER WENT HOME THAT NIGHT, kissed his wife, played with the dog, and petted the cat. Then he told Allison the news about Kojima. She was nearly as surprised as he'd been.

"You're kidding. He's really coming to L.A.? Do you think he'll try to contact you?" she asked, half hoping that he wouldn't. Newcomer was already working a case that worried her. Allison had only recently learned that the infiltrated roller-pigeon group included gang members. She tried not to think what might happen if they ever discovered he was carrying a gun and wearing a wire.

"No, I don't think so. Who knows if he'll even be at the fair? I'm just going to check it out, but I doubt that anything will happen," he assured her.

Maybe not, but Newcomer was counting the days until the weekend. He had immediately contacted U.S. Immigration and Customs Enforcement special agent Jamie Holt to learn that Kojima had already entered the States. Kojima hadn't been put through a secondary inspection this time. Instead, he had walked straight through Customs.

Newcomer now fell asleep every night and woke every morning with Kojima on his mind. It seemed as if he saw butterflies everywhere again. He took it as a sign.

He woke early Saturday morning, showered, and slipped into freshly pressed jeans and a clean polo shirt. The only vestige of his roller-pigeon persona was his long, droopy Village People mustache. He opened his bureau drawer and placed his wedding band on a pile of neatly folded socks. Then he stared at his reflection in the mirror. Ted Nelson, the eager young butterfly dealer, was beginning to reemerge.

Newcomer arrived at the Los Angeles County Museum of Natural

History before the place was even open. Thirty minutes had never seemed so long as he waited for the time to pass. The museum doors were promptly flung wide at 10 AM. He didn't waste a moment but quickly made his way inside.

The entrance hall yawned before him like a giant maw from which he could have sworn there was a seductive siren's call. Looming in the center was the skeleton of a large *T. rex* that seemed to magically transform into a samurai warrior suited up for battle. Newcomer grabbed a vendor's map and hurried down the hall, hoping to break the spell. It took less than a minute before he spotted Yoshi Kojima standing at the booth of a friend. He had a wooden box for pinning and transporting butterflies under one arm. Newcomer could only imagine what treasures it held.

He stood back and watched Kojima flit from booth to booth as serious collectors gathered around him like bees drawn to honey. He was the undeniable belle of the fair, the king of butterfly smugglers. He could provide collectors with their greatest desire, perfectly mounted specimens of winged enchantment.

Even so, there was definitely something different about Kojima. Newcomer gradually realized what it was. The man had been more vibrant and robust the last time that he'd been here. Now he appeared thinner and frailer.

Or maybe I'm just feeling better about myself as an agent, Newcomer mused.

He followed from a safe distance, waiting for just the right opportunity to approach. The museum was a tough place to manipulate an encounter, and Newcomer hoped to make it appear as though they'd accidentally run into each other.

He studied the long hall that ran from wing to wing. The passageway narrowed at one point, and Newcomer stationed himself near there, a hunter stalking his prey. Kojima was bound to pass through on his way to the other exhibits. Newcomer's patience paid off as Kojima finally approached. Newcomer nonchalantly walked toward him with perfect timing. They bumped into each other in the narrowest part of the hallway.

Newcomer smiled at Kojima, though Kojima appeared not to know who he was at first. Then Kojima slowly recognized the young man standing before him.

"Yoshi! How are you doing?" Newcomer eagerly inquired, as though the two had parted on good terms.

"Ah, yes. Ted, isn't it?" Kojima responded hesitantly while shifting his weight from side to side. "Are you still selling butterflies?"

This was the moment Newcomer had been waiting for. He'd repeated the mantra so often that he had the timing down pat. "Yeah. Man, things are going great. Only I'm not selling on eBay anymore."

"Oh, really?" Kojima asked, his interest perking up.

Got him, Newcomer thought as he began to weave his story.

"No, I'm doing something better. I've developed a steady customer base. I decided to sell to a select group of customers just like you do, Yoshi. You were right all along. That's definitely the way to go," Newcomer said, gently stroking his mentor's ego. "I'm very careful about new people. I developed a private Web site that requires a password to enter."

Kojima's shoulders relaxed from where they'd been bunched up around his neck. Perhaps this wasn't a confrontation, as he had feared. Kojima assessed the young man before him and then asked the question Newcomer had hoped to hear. "Who's your supplier?"

Newcomer was tempted to pinch himself just to make sure he wasn't dreaming. His words came easily. After all, he'd rehearsed them in his sleep for years. "I can't tell you, of course. But it's this guy in L.A. who has his own boat. He travels all over the South Pacific twice a year and gets whatever butterflies I want. The problem is I never know when he's coming back. He's also not very good at picking material. A lot of the stuff I end up with is spotty."

Kojima gave an understanding nod. Of course, what else could Nelson expect? There was no other dealer like Yoshi Kojima.

Newcomer delivered what he considered to be the coup de grâce. "You don't know it, but I've got something else to thank you for, Yoshi. You saved my life."

"I did?" Kojima responded with a puzzled expression.

Every cell in Newcomer's body tingled as he moved in for the kill. "Yeah, man. You're not going to believe this, but some jackass turned me in to Fish and Game a couple of years ago."

Kojima's eyes grew wide as an owl's, and he quickly glanced around as if to seek refuge. His intuition had been correct. This was a trap after all. "Oh, really?" he squeaked, his voice having risen an octave and grown taut as a rubber band.

Newcomer relished the moment. "Yeah. I don't know who it was, but some jerk called the tip line. Wouldn't you know? A couple of Fish and Wildlife agents came to my house. It was just like what happened to you. These bozos knocked on my door and demanded to come in and look at my stuff. Thank God, you gave me good advice. I did exactly as you'd told me. All my illegal butterflies are stashed inside a warehouse. I remembered how you'd handled that Fish and Wildlife agent and let the joker in. Thanks to you, they didn't find anything and had to leave me alone."

Kojima's clenched fists began to loosen. The boy apparently knew nothing. He broke into a smile as Newcomer continued to pour it on.

"I owe it all to you, man. The whole eBay thing was originally your idea. How to keep Fish and Wildlife away was your idea. Why, if it wasn't for you, I might be in jail today," Newcomer said, adding the final touch. God, revenge was sweet, and he was only just beginning.

Kojima's ego was now pumped up to the size of a hot-air balloon. Any more flattery and his feet would have left the ground.

"Kojima was clearly thinking, 'This guy Ted thinks I'm the coolest dude around,'" Newcomer recalled. "He also figured, 'Wow! Ted's still dealing in CITES butterflies and hasn't been caught. Maybe he's someone worth dealing with after all.'"

Kojima told Newcomer that he was staying with his father-in-law but was in L.A. for the weekend. They agreed to meet later that day for lunch.

Newcomer could barely believe this turn of fortune. His cover was still secure after all these years. A cat might have nine lives, but Newcomer had at least three of them. He paced the halls and imagined what Marie Palladini was going to say when he told her. He also daydreamed about his upcoming lunch with Kojima. He was half afraid that Kojima would change his mind and not appear.

Kojima met him at one o'clock on the dot. "I'll take you to a good place for Korean barbecue. It's where I used to go when I lived in L.A. You'll like it a lot," Kojima proposed.

Newcomer didn't dare dissent, though he was now vegetarian and eating meat was the last thing he wanted to do.

"By the way, I didn't rent a car. You don't mind driving, do you?" Kojima asked.

Newcomer ran through a quick mental checklist. As far as he knew there was nothing in his vehicle to give him away. They got into the Tahoe and took off as Kojima guided him toward lower Hollywood.

"I've been very sick, you know," Kojima offered of his own accord.

"You have? What's the matter?" Newcomer asked. Maybe that's why Kojima had seemed to disappear for the past few years.

Kojima explained that he'd had bypass surgery after suffering a heart attack in Japan. Pulling open his shirt, he showed Newcomer the long angry scar that ran down the front of his chest.

"You see? I'm not beautiful anymore," he said with a rueful grin.

Newcomer laughed lightly. "But you're all right now?"

"Yes, I'm okay, but I had to stop selling butterflies for a while. My son helped with my insect business while I recovered." Kojima pulled out his cell phone and produced a digital photo of a young man who appeared to be about twenty-five years old and Eurasian. "That's my boy, Ken."

"He's a good-looking guy. Is he in the insect business, too?" Newcomer asked, wondering if he should also focus on Kojima's son.

"He helps me sometimes, but my boy works for National Geographic now. I taught him everything he knows," Kojima proudly related.

They pulled up to a hole-in-the-wall restaurant that probably should have been condemned by the Board of Health. The smell of rank greasy meat assaulted Newcomer's nose before he'd even stepped out of the Tahoe.

"Why don't you get a table? I want to make a pit stop in the bathroom," Newcomer said as they walked inside.

He waited until Kojima was no longer in sight and then dashed back out to his vehicle.

Please, let it be in here, he prayed while rummaging through the glove compartment. *Yes!* He sighed in relief as his fingers wrapped around the small recorder. He turned the machine on and hid it in his pocket. Then Newcomer walked back inside and joined Kojima at the table.

Kojima ordered lunch, and Newcomer's stomach lurched as a plate of mystery meat appeared before him.

Kojima immediately launched into the topic of butterflies. "There's a good market for them in both Europe and Japan, but European customers are very scared these days."

"Oh, yeah? Why is that?" Newcomer asked, subtly pushing his food around on the plate.

"They're afraid they'll get caught just like you almost did," Kojima replied with a laugh. "What type of butterflies are you selling now anyway? Mostly birdwings?"

Bingo, Newcomer rejoiced. Kojima had just given him the perfect opening. Named for their exceptional size and beautiful birdlike flight, birdwing butterflies are protected under CITES.

"Yeah, everybody that I know wants those. In fact, I'm looking for some *Ornithoptera chimera* at the moment."

An iridescent green, the butterfly's wings have streaks of black and are bedecked with panes of gold. They look as if a glassblower had painstakingly created them.

"I heard the dealer at Insect International in Texas sells them for $236 a pair," Newcomer nonchalantly added, hoping that Kojima would react.

He got his wish and then some as Kojima waved his hand in disgust. "Don't buy from that guy. He carries the ones from Indonesia. The really good specimens come from Papua New Guinea and are very hard to get. I have ten pair of those and can sell them to you for just seventy dollars a pair. I also have Victorias from the Solomon Islands. They're very expensive, but I'll give them to you for forty dollars each."

CITES butterflies danced like sugarplums in Newcomer's head, each more exotic, gorgeous, and rare than the next. Man, this was like the good old days when Kojima was undercutting him on InsectNet. No wonder all the other dealers hated him.

"I can't believe how many new contacts I made at the bug fair this morning. So many people came up to me, all wanting to buy my butterflies. And that's not counting all the people that already knew me there," Kojima bragged.

"That's because you haven't been here for so long. Last year everybody at the fair asked, '*Where's Yoshi? Where's Yoshi?*' They missed you because you have the best stuff by far," Newcomer said, heaping on the praise.

"I also have the cheapest," Kojima sagely added. "Oh, I'm so happy today. I made almost twelve thousand dollars in only one hour."

"Wow, Yoshi! That's terrific," Newcomer responded. *He's got to be kidding. What the hell had he sold to reach that price so quickly? No wonder the guy could afford to fly all over the world.* "Listen, I'd be interested in buying those *chimeras* from you. Will they come with CITES permits?"

Kojima chewed on a piece of gray meat as if it were a wad of gum while shaking his head. "There are no permits, but it doesn't matter. So many of them are fake anyway. My friend knows a guy in Indonesia that makes them all the time."

"Oh, yeah? But don't Appendix II butterflies need CITES permits to be legally imported into the U.S.?" Newcomer asked, wanting to get Kojima's admission on tape.

Kojima spit out a piece of fat. "Such a stupid law. I'll send them to you by Express Mail. Don't worry."

"But won't Customs check?" Newcomer persisted.

"No, they never look at Express Mail," Kojima confirmed.

Newcomer knew Kojima was correct. Too much mail flooded into the United States for every package and parcel to be opened. The mail was a notorious black hole through which almost anything illegal could be sent.

"Great. I already know a buyer who will pay a hundred bucks for a pair of *chimeras*. But don't you ever have trouble importing butterflies into Japan without CITES permits?" Newcomer prodded, curious as to Kojima's smuggling tricks.

"No, it's very easy. The only punishment in Japan for getting caught is that you have to write you'll never do it again on a piece of paper," Kojima explained. "The one difficult thing to import is Appendix I species. But I don't mind. I never have trouble with that, either. In fact, I can get Alexandras, too."

Alexandras.

The word shot through Newcomer's brain like a five-alarm fire. Queen Alexandras are the biggest butterflies in the world and the flagship emblem of butterfly conservation. Their wingspan can run up to a foot, making them larger than many birds.

Lord Walter Rothschild had named the species for Queen Alexandra of England in 1907. The butterflies flying in Papua New Guinea at that time must have been a sight to behold. They're so big that early explorers knocked them out of the sky with their shotguns. An arresting blue and green, the male has a deep yellow abdomen that resembles a small bar of gold. The females dwarf their male companions with large chocolate-brown wings dotted with cream while a vibrant red tuft of fur shines bright as a stoplight on their thorax.

Queen Alexandras aren't seen very much these days except for those that are caught, killed, and mounted. Their small strip of coastal rainforest continues to be destroyed to make way for oil palm, cocoa, and rubber plantations. The butterfly's existence is so precarious they've been listed as endangered. It's illegal to capture and sell specimens, yet here was Kojima boasting that he could obtain them.

"Can you really get those?" Newcomer asked, half appalled and half fascinated.

"Of course," Kojima said. "You just have to be smart, have the right contacts, and know exactly what to do. I have them shipped from Papua New Guinea to my friend in Czechoslovakia. He wraps them in newspaper and writes the scientific name of a moth on the box. Japanese Customs doesn't suspect anything illegal coming from Europe and so never bothers to check."

Kojima was the real deal, the gatekeeper to all that was elusive and desired in the butterfly world.

"Those must be pretty expensive butterflies. Who do you sell them to?" Newcomer probed. He was thrilled to be getting the information on tape but still couldn't quite believe it.

"The Japanese buy them. I pay seven hundred and sell them for eight thousand dollars apiece. That's for the male. Females are more expensive because they're much bigger," Kojima confided.

"What about over here? Do you ever sell Alexandras to people in the U.S.?" Newcomer asked. He figured he might as well hit while Kojima was willing to talk.

Kojima looked at Newcomer's untouched plate. Newcomer quickly speared a piece of meat and gobbled it down. "Sometimes. But there's no reason to do it here when I sell a lot of them in Japan." He stopped eating and stared at Newcomer for a moment. "Tell me, why did you grow a mustache?"

Newcomer's stomach instantly flip-flopped. He wasn't sure if it was the meat or Kojima's abrupt question. "Oh, I was just in the mood for a change."

Kojima nodded and then smiled. "I like it a lot. You're such a handsome guy, just like my actor friend here in L.A. He's in a police show on TV all the time."

Newcomer was relieved but not yet ready to change the subject. "I wonder if any of my customers would buy an Alexandra."

Kojima looked at him sharply. "You better be careful."

Take a step back, Newcomer cautioned himself. *Don't push, or you're going to blow it.* "Yeah, I don't think they would. They're probably too nervous."

"Besides, I don't want it to get back to me," Kojima testily warned.

"Don't worry about that. I've got things pretty well covered," Newcomer tried to assure him.

"Oh, really?" Kojima appeared to be amused as he put on his glasses and picked up the check.

"Hey, you're getting old on me if you need glasses," Newcomer gently teased.

Kojima studied him with what appeared to be a touch of nostalgia. "You're such a young boy." He'd had no worries once. Someday Ted Nelson would understand.

Newcomer couldn't help but be flattered. At the same time, it felt strange to see himself through Kojima's eyes. "Oh, yeah? How old do you think I am?"

Kojima looked closely at him. "Thirty-three, I think."

"Uh-uh," Newcomer responded with a sly smile. Sometimes his youthful looks played to his advantage.

Kojima leaned in toward him. The boy didn't have a line on his face. "Thirty-four or thirty-five at the most."

"I'm going to be forty," Newcomer told him.

Kojima looked surprised. "No, really? You still look so nice." How odd. He was usually very good at guessing a young man's age.

Enough with the small talk. Newcomer brought the conversation back around to the matter at hand. "So, how are we going to do this butterfly thing, anyway?"

Kojima placed his palms on the table with fingers spread wide like two resting butterflies. Perhaps he'd learned his lesson. He'd give the boy another chance. "I'll send you a few at first. If no problem, then I'll send more."

I can promise there won't be one, Newcomer silently vowed. His plan had been well thought out.

"That sounds good. I guess you just have to make sure not to put your name on it. After all, I don't want Fish and Wildlife to come sniffing around," Newcomer replied.

Kojima leaned back and smiled. "My name is no problem for me. They cannot touch me outside the United States."

Newcomer got the gist. If Ted Nelson wanted to get involved, then Ted Nelson took all the risk. Newcomer had to hand it to the man. Kojima certainly knew his stuff. He decided now was the time to ask about something that puzzled him.

"Hey, remember you once told me that you have two passports? American and Japanese?" Newcomer mentioned nonchalantly.

Kojima nodded. "Yes, I got my American passport in 1984."

"But didn't it expire after ten years?" *How in the hell did he manage to pull this stuff off?* Newcomer was determined to get to the bottom of it.

"No, I continue to renew it," Kojima informed him.

"Oh, so then is it the same name on both of them?"

Come on, come on. Just spill a little more information. Kojima had been so open about endangered butterflies that maybe he'd decide to talk about this too.

Kojima wasn't that easily tricked. He shook his head and laughed. "That's my secret. My American passport has an American name on it.

Also, I never carry butterflies. That's why Fish and Wildlife can't catch me, because there's no evidence."

"Then how did you get the butterflies here to the fair?" Newcomer asked. The man was a walking encyclopedia of how to successfully smuggle, and he wanted to learn all that he could.

"My son brought them in for me. He doesn't have my last name, so Customs never checks him," Kojima revealed.

Newcomer let loose a low whistle. Damn, Kojima was good. "You're very slick, Yoshi. You're definitely the king of butterfly smugglers. Did your son fly into Los Angeles?"

Kojima zeroed in on him, and for a moment Newcomer felt uncomfortable. What did Kojima see when he looked at him so intently? "Yes, he did. Tomorrow you bring me the list of butterflies you want. Meet me at the fair at four PM, and then we'll go to dinner. Now I want you to take me somewhere," Kojima said.

The two men left the restaurant and got into Newcomer's Tahoe.

"Where to?" Newcomer asked, wondering what Kojima's next stop would be. Perhaps to visit a client or something else that Newcomer would find useful.

"To a bathhouse not far from here," Kojima instructed.

It was the last thing Newcomer expected to hear as he drove further into Hollywood. The request seemed odd, but then maybe Kojima was looking for a quick sexual encounter. On the other hand, lots of Asian men liked steam baths. It could also be perfectly innocent.

"Hey, we're in Hollywood. Maybe we'll see a movie star while we're here," Newcomer mused with a laugh. It was a game that he and his wife liked to play since moving to L.A. They'd always try to spot a celebrity to add to their list.

"What movie stars do you like?" Kojima asked casually.

Newcomer mentioned one whose film he'd recently seen. He quickly learned Kojima believed that every actor in Hollywood was gay. Newcomer couldn't name one celebrity without being told the "true story" of their homosexuality. It was as if Kojima were a walking, talking tabloid.

"No way you can tell me that this guy is gay," Newcomer challenged and gave the name of a well-known Hollywood stud.

"Sometimes you have to suck a few pee-pees to get ahead in Hollywood," Kojima joked, enjoying the game.

"Well, then, how about your actor friend?" Newcomer asked, bemused.

"Oh, he's so bi," Kojima replied with a laugh. "He likes both women and men. It's because he has such a big one. He's big as a horse!"

"Okay, I give up. You win. So, what about this bathhouse, anyway? Are you going to get a happy ending there?" Newcomer teased.

"Oh, no. For Japanese men, this is very common in our culture. I go to the spa every day. I don't need special ones anymore. But they'll do it for you, if you like," Kojima offered.

Newcomer had no doubt. "You've got to be careful with those special spas. You could end up going to jail for that," Newcomer warned.

"I know all about the laws," Kojima informed him.

Little did he know, Kojima was telling the truth. Newcomer had yet to learn that Kojima had twice been arrested in Los Angeles for obscene live conduct and soliciting a lewd act back in the 1980s.

Newcomer had had enough talk of gay movie stars and bathhouses. It was time to get down to business. "Anyway, all my clients want CITES stuff. Sometimes I can get permits and sometimes I can't. But I'd be in big trouble if Fish and Game ever caught me."

"Don't worry. There's no need for so much CITES stuff. You can sell different butterflies and still make lots of money," Kojima assured him. "Just charge little enough so that you sell everything quickly. That's the Japanese business way."

Perhaps so, but it wasn't the information that Newcomer had been fishing for. *Take it easy. Don't jump the gun and make him hinky,* Newcomer reminded himself.

Kojima pointed to the bathhouse as they approached. "You want to come in with me? They have a very nice sauna."

"Not today, but as long as there are pretty women inside, then it's a good place to be," Newcomer replied, eager to drop off Kojima and leave.

"Right now there are lots of kinky guys that come also," Kojima revealed.

That was no surprise. Hollywood was filled with all kinds of people. Still, was Kojima talking about S&M or something else?

"Are you referring to gay guys?" Newcomer asked, unsure exactly what he meant.

"Yes. Sometimes they suckee right in front of me. That's okay." Kojima said, and turned to him with a smile.

So that was it. Kojima was a voyeur, or perhaps he was bisexual. It made no difference to Newcomer. He had his eye on the prize and wanted to keep it that way. After all, a new level of trust had just been established between them. Kojima had put him through the fire by turning Ted in, and he had emerged unscathed. The young novice had proven himself.

Newcomer genially laughed. "In that case, have fun in there. I'll see you tomorrow after the fair."

He watched as Kojima walked into the bathhouse. Newcomer had blown the case the first two times. He didn't plan to do it again. He took a deep breath and fully submerged to once more become Ted Nelson the butterfly dealer.

Newcomer was back in the game; this time he intended to win.

The Game Is On

I want to be with those that know secret things or else alone.
—Rainer Maria Rilke

Allison was waiting for Ed when he got home that Saturday night. Operation High Roller had put a temporary glitch in their schedule, but she hoped the case would soon be over and their lives could return to normal.

Newcomer walked in the door, and any thoughts she'd had of normalcy were promptly shattered. He'd left their apartment at nine that morning. By the time he came home, the Kojima case was not only back on but had returned with a vengeance. Newcomer was more obsessed with catching Kojima than ever.

The news of Kojima's revival blindsided her. "That's great. But does this mean you're going to be working both undercover cases at the same time?" she asked.

Newcomer thought about it a moment. He and Kojima hadn't gelled the first time around. Now their chemistry seemed to be cooking. He didn't intend to give up either case. He would just have to learn to juggle them both somehow. "Don't worry. It will all work out," he assured her, but even he wasn't certain he could keep his promise. Allison placed importance on a close relationship, and Ed was already distracted. It didn't bode well.

He immediately took a shower just as he always did after seeing

Kojima. It was the only way that he could feel clean again. Then he promptly went to bed.

He spent Sunday morning creating a wish list of butterflies. Included in the roll call were Appendix II species, along with an Appendix I butterfly that was considered impossible to get—the Queen Alexandra. Kojima had mentioned it, and Newcomer figured he might as well go for the prize. It was the best way to learn the extent of Kojima's network.

He arrived at the insect fair at four o'clock sharp and Kojima met him at the appointed spot.

"Come. Let's go over here. I want to show you something," Kojima said with a gesture. He led Newcomer away from the crowd to a quiet corner of the museum. He made sure no one was watching and then pulled a small plastic box from his fanny pack.

"What's that?" Newcomer asked, secretly hoping it wasn't another live bug that he'd have to deal with.

Kojima held the box up for closer inspection.

Newcomer gazed at what appeared to be five small brown pieces of turd.

"These are live cocoons of *Papilio indra kaibabensis*," Kojima reverently whispered.

"You're kidding! Where did you get them?" Newcomer inquired, taking a better look. This was exactly what Mendoza had tried to arrest him for back in the nineties.

"My friend collected them in Grand Canyon National Park. He traded them for some butterflies that I brought," Kojima said.

"That's really risky. He could get in big trouble if he's ever caught," Newcomer warned, wondering who his friend was. If he was selling cocoons to Kojima, then he was probably selling to others, as well.

Kojima nodded in agreement. "The butterflies are found mostly in the park, but some are collected outside. That makes it difficult to prove where they came from."

Kojima wasn't kidding. That's how people like his friend got away with the ruse. Even most collectors couldn't tell if the *P. indra kaibabensis* were harvested from inside or outside the park's boundary.

Kojima planned to take the cocoons back to Japan, where he would hatch, kill, and mount them. They were worth only a few dollars in the United States but could be sold for five hundred apiece to Japanese collectors. They'd pay a fortune for butterflies that came from the American Southwest. "American butterflies are becoming very popular in Japan. You could help by collecting the larvae, letting them cocoon, and trading or selling them to me," Kojima suggested.

Thanks, Yoshi. You just gave me a great idea.

"Sounds good to me," Newcomer agreed. Maybe he could meet Kojima's friend that way and go collecting with him. Then one more violator would be caught in his web.

Kojima produced a thick wad of cash from the fair and fanned the bills like a gambler.

"See how well I've done today? Now I'll take you to dinner," Kojima said, and stuffed the money back in his fanny pack.

Enjoy it now, Newcomer thought grimly. *You won't have much use for it where you're headed.*

He directed Newcomer to drive to Beverly Hills. Kojima's choice of restaurants was again a carnivore's delight. Newcomer's stomach turned as they pulled in to a Lawry's steakhouse.

They ordered dinner, then Kojima asked to see the list of butterflies that Nelson had drawn up.

"I have a few customers who would be interested in getting the rarest CITES butterflies," Newcomer said.

Please don't let me have blown it by doing this, he prayed as Kojima perused his typed request.

There was *Ornithoptera chimaera* brandishing green and gold as precious as set emeralds, along with the brilliant blue *Morpho cypris,* that flaunted a broad band of white on its wings. The neotropical *Morpho rhetenor helena* shimmered in shades of blue and green foil. The butterfly was aptly named after Helen of Troy, while the word "*Morpho*" is an epithet of Aphrodite and Venus.

Ornithoptera goliath atlas was second in size only to another butterfly on Newcomer's wish list, *Ornithoptera alexandrae,* Queen Alexandra's birdwing. Finally, there was *Parnassius autocrator,* one of the most highly desired butterflies in the world, found only in the Hindu Kush of Afghanistan and the Pamir mountains of Tajikistan.

"It's no problem. I can supply them all," Kojima replied nonchalantly as he cut into his steak. In fact, he already had some of the butterflies at home in Kyoto. The rest could be obtained through his suppliers in Indonesia and Papua New Guinea.

"If you like, I can also get you *homerus* from Jamaica," he added casually, as though it was an afterthought.

Either Kojima's the biggest BS artist around, or he's got one major pair of cojones, Newcomer reflected in stunned silence.

If Queen Alexandra is the Holy Grail, then *Papilio homerus* is the Golden Fleece. It's the largest butterfly in the Western Hemisphere and considered a true collector's prize. The velvety black giant swallowtail is marked with broad bands of gold, making it as gorgeous as any Southeast Asian butterfly. Silvery blue fireworks erupt on its fluted hind wings before delicately tapering off into elongated teardrop-shaped tails.

Papilio homerus lives exclusively in Jamaica and is such a symbol of national pride that it's found on the country's one-thousand-dollar bank note. Making it all the more alluring is that it is extremely rare and has no close relatives anywhere else in the world. Even so, *P. homerus*'s habitat has been pared down to just two island sites—the montane forests of the Blue Mountains and the impenetrable Cockpit Country. But even the remoteness of these two wild regions can't help to protect them.

Named after the epic poet Homer, the butterfly is also known as "the Traveler" because it flies as languidly as a bird. It soars seventy feet high in the forest canopy but descends from the treetops in July and August to sip at puddles on the forest floor. That's when collectors sneak in and illegally catch them.

"I can supply Alexandra and *homerus* in limited quantity," Kojima continued. "Now, what else would you like?" Kojima could already tell that the boy was impressed, just as he should be.

Maybe it was reckless, but Newcomer decided to throw caution to the wind and go for the Top Four hit parade. Kojima had opened the door, and time was of the essence when dealing with him. He never knew when the man might reverse course and turn tail. "Well, I've got one customer who's particularly interested in the Grand Slam," he ventured.

"What's that?" Kojima asked, seeming to be momentarily puzzled.

"It's the Big Four," Newcomer replied, beginning to feel more secure. "You know, *Alexandrae, homerus, chikae,* and *hospiton.*"

Kojima had already mentioned two of the species. He might as well go for the gold. They're the four most endangered butterflies in the world and are all CITES Appendix I. Newcomer held his breath.

"Oh, I can get them," Kojima replied, as though they were nothing more than goldfish.

Newcomer couldn't believe what he'd just heard. He hadn't really expected Kojima to pony up. Not even museums could get those species, since they were simply not around. Kojima was either being inexplicably rash or was supremely confident that he'd never get caught while living in Japan. Or maybe he was just becoming downright greedy.

Newcomer decided to play it safe and not seem too eager at first. "The only thing is, I'm really nervous about doing it," he said.

"You just have to be very careful. Don't accept any checks. Make sure it's cash only, and always get half the money up front," Kojima advised. "Also, never let the buyer go to your home. Always meet him somewhere else, like a parking lot or a grocery store." He provided explicit instructions on how to avoid a paper trail. No wonder he'd gotten away with it for so long. Kojima took care that no evidence would lead back to him.

"Don't worry. I'll send a few trial packages by Express Mail to make sure they aren't intercepted. If all goes well, then I'll send the rest," he promised. Kojima would mark the contents of each box with scientific names of unprotected species as an added precaution. "That way, it won't be a problem even if we're caught," he explained.

Kojima would already be safely home in Japan, and Ted Nelson could deny any knowledge of the contents. After all, if the butterflies were listed as legal species, then how could Nelson know that something else was packed inside?

Newcomer now felt emboldened enough to ask the question that had haunted him for the past three years. "Yoshi, there's something I have to know. Why didn't you ever send me any butterflies when we first met?"

For a man who was forty years old, Ted Nelson still seemed like a boy in so many ways. "Because you were too hard to contact," Kojima bluntly told him. "But we'll do it differently this time. You should sign up for a Skype account. That way, we can talk for free on the phone when I'm back in Japan. It will be much easier."

Skype is an Internet-based phone service that lets you talk to anyone in the world. It's compatible with video Internet cameras and allows people to see and speak to each other in real time. Digital files can also be sent and received. It was a new company, and Newcomer had never heard of it.

"If you get a webcam, we can see each other and I can show you what butterflies are available. That way you can tell me what you want to buy," Kojima suggested.

Newcomer couldn't have been any more pleased if he'd dreamt up the plan. It was the perfect way to collect evidence. He'd be able to record everything Kojima said, along with every butterfly that Ted

Nelson was offered. It seemed almost too good to be true. Kojima had just handed himself to Newcomer on a silver platter. "What a great idea! That would be terrific." Newcomer profusely thanked him.

Kojima ended dinner with what were becoming his usual quips about sex, gays, and heterosexuals. Everything was just vague enough to furrow Newcomer's brow and keep him guessing. *Exactly what's going on?* he wondered. *Is this guy a homophobe or possibly a closet queen?*

Kojima's remarks were purposely nebulous to be denied either way, and Newcomer felt no need to comment. It made no difference to him. All that mattered was that they were back in business together, and the case was shaping up to be better than ever.

NEWCOMER POPPED INTO MARIE PALLADINI'S OFFICE bright and early on Monday morning.

"Hey, guess what? I spent the weekend hanging with Kojima. Apparently, we're pals again," he said with his trademark boyish grin.

Palladini was as astounded as Newcomer had been.

"You're not going to believe it, Marie. We're back in business big time. Kojima is offering to sell me Appendix I and Appendix II butterflies. The guy's got an unbelievable worldwide network and can get endangered butterflies from absolutely anywhere. We're talking about butterflies that you don't even see in museums. I'm hoping to learn who his suppliers are and snag a list of his clients."

He told her about Skype and how his plan would work. There was just one catch. He needed "buy" money to pull it off. It wasn't a simple request. Palladini would have to plead her case to the special agent in charge of the region, Paul Chang, and hope for the best.

Getting money for an undercover case should be easy; instead, it's like pulling teeth. An agent needs to be persistent and have a good sales pitch. Fortunately for Newcomer, Palladini had years of credibility. Had a less experienced agent made the appeal, it most likely would have been turned down. The usual response tended to be "Don't you guys have other priorities to deal with?"

To her credit, Palladini had chops. She also had a bit of luck. Paul Chang was a butterfly buff and realized the importance of the case.

The next thing on Newcomer's to-do list was to call the assistant U.S. attorney (AUSA) and advise him that the Kojima case was back on.

In the good old, bad old days, there'd been no special unit at the Los Angeles U.S Attorney's Office to whom Fish and Wildlife agents could present their cases. They'd had to trudge to the Criminal Complaints section and take a number as if they were at a deli. It was there that a revolving group of new attorneys decided whether a case lived or died. More often than not, wildlife cases were automatically declined for prosecution due to being viewed as inconsequential.

"We had a tough time. We begged, borrowed, and stole, pretty much," Palladini recalls. "I'd go and wait my turn in the room to see a U.S. attorney just like everyone else. When I finally got in, I'd give the attorney my affidavit only to have them say, 'Now, what exactly is the Lacey Act?'"

They had no idea what Palladini was talking about. She'd have to explain Fish and Wildlife laws to a new attorney each time. That was the least of her problem. The L.A. office was busy with other crimes, which took precedence. There were murders, gunrunning, and drug cases involving the Colombian cartel. FWS agents found themselves competing against the DEA, the IRS, and the FBI for attention. Wildlife cases didn't stand a chance. No one understood them, or cared.

Sam Jojola remembered wooing assistant U.S. attorneys by taking them to lunch and for boat trips just to try and get his cases heard. It didn't matter that USFWS agents were federal employees. They were reduced to acting as lobbyists to further their own cause.

"I'd give them a Marie Callender's cherry or blueberry pie whenever they agreed to sign an indictment," Jojola recalled. That always helped to move his cases along.

Bill Carter was the one assistant U.S. attorney in Los Angeles who was sympathetic to Fish and Wildlife's plight. He took on as many environmental cases as possible. Carter soon found himself battling pollution by oceangoing cruise ships, illegal disposal of hazardous waste, and smuggling of ozone-depleting substances. He tried to fit in the few wildlife cases he could, but it proved to be too much for a one-man band. Carter asked environmental attorney Joe Johns to join him in 1998.

Johns was a gung ho, do-or-die attorney whose first calling had been to join the military as a Green Beret. He turned to law only after his parents put the kibosh on that decision. Johns started his career at the San Bernardino County District Attorney's Office and soon began to run its Environmental Crimes Unit. The atmosphere for environmental enforcement at that time was as raucous as the Wild, Wild West. It was so much fun that Johns set his sights on moving up the ladder and, five years later, decided to become a federal prosecutor. He was offered the job of assistant U.S. attorney for the Eastern District of California in Fresno, where he once again worked with Environmental Crimes.

Johns casually broke the news to his wife.

She'd been down this road before and took the news in stride. "You've got to be fucking kidding me. You want to take me from the armpit of Southern California to the armpit of central California?"

They packed up the kids and moved. Johns worked in Fresno for two years and, in the process, built a name for himself. He became the bane of ranchers and farmers in the district for upholding environmental laws. He did so well that he was referred to as an Aryan fascist tree hugger.

Johns received a call from the Department of Justice in Washington, D.C., in 1997. Janet Reno, then attorney general, had a new pet project that she wanted to get off the ground, the Joint Center for Strategic Environmental Enforcement. The task force was an off-the-books intelligence outfit whose mission was to learn which companies, or industries, were likely to be breaking environmental laws. The goal was to generate leads for environmental crimes by using cutting-edge high technology.

Johns was invited to become the center's director for a three-year period. He jumped at the opportunity, only to have things quickly fall apart. A western senator learned of the group's existence, and a firestorm erupted after just the first year. The task force was called everything from the Green Gestapo to the Nazi Environmental Fascist Police. The term "Posse Comitatus" was even bandied about. Finance Committee hearings were held on whether the group should receive government funding. The nays outweighed the yeas, funding was pulled, and the Joint Center for Strategic Environmental Enforcement was immediately shut down.

For the first time, Johns found himself without a job. He momentarily considered going into private practice and becoming a hired gun for whoever was willing to pay. The work wouldn't be as satisfying, but the money would certainly be better. That was when Bill Carter approached with an offer that he couldn't refuse. Why didn't Johns join him in L.A. and they'd take on environmental crimes together at the U.S. Attorney's Office? Joe Johns's idealism won out.

The two men worked together in the Public Integrity unit, investigating and prosecuting government fraud, but they had a much larger vision in mind. They swiftly built up a track record with which to approach the California attorney general. Carter and Johns wanted to open their own unit solely devoted to prosecuting environmental crimes. The Environmental Crimes section was born in January 2000. Its birth couldn't have been timelier. Wildlife crime was on an upswing, and what was being caught was just the tip of the iceberg.

Newcomer first approached Bill Carter when the Kojima case began back in May 2003. Carter had been less than enthusiastic at the time. The case hadn't seemed very exciting to him. He eventually handed it over to Joe Johns. Newcomer felt sure the case was doomed when he heard the news.

"I thought, 'Oh, shit! Joe's not going to go for this. It's not his kind of case. Butterflies aren't hard-charging enough for him.'"

Johns is a hard-core adrenaline junkie who spends weekends doing serious backpacking and rock climbing. His idea of an awesome time is a long-distance hike alone in the wilderness.

"I hike at night knowing darn well there are cougars out there, but I love the adventure," Johns enthuses. Tall and lean in his Brooks Brothers suit, he's a cross between a gangly hyperactive kid and wild man Davy Crockett. "You're super-alive because you're on total red alert. Every twig that snaps, every raccoon moving through the chaparral, sounds like an elephant on your tail. I don't think you're ever quite as awake as when you're running on the ragged edge."

Johns is a lot like Newcomer. He's not so much into creepy-crawly bugs. His interest lies in tigers, bears, and megafauna. Newcomer worried that Johns wouldn't build up a head of steam over tiny butterflies. The federal court system is also very discriminating about the types of cases they charge. They only want to prosecute the biggest and the best. U.S. Attorneys' Offices don't like to lose. The tendency is to turn down anything that's small and not high-profile.

Newcomer gave Johns his best sales pitch. "Kojima is the world's most wanted butterfly smuggler. We've tried to catch him a number of times before and never have. He's always gotten away, but I think that I can get him this time."

Joe Johns found himself intrigued. He'd worked on everything from illegal caviar to smugglers walking through Customs with pygmy monkeys stuffed down their pants. This case was something new. Besides, Johns loves a good fight. It's one of the reasons he went into law. He's an aggressive hard-charging prosecutor who considers himself a legal gladiator, a Green Beret warrior for the environment. To Johns, a trial is little more than a toned-down courtroom bare-knuckle fistfight.

He agreed to work on the case and quickly gained a vested interest. Traditionally, the first time an agent ever sees an assistant U.S. attorney is upon walking into the attorney's office with all the evidence.

"They'll hand it to me and say, *here you go.* That's old-school. I've never operated that way," Johns explains. "Agents may think they know it all. But guess what? They've got a law-enforcement approach, and I've got a prosecutorial one. If we put our heads together, we're going to come up with a much better game plan."

Newcomer was that unusual combination of a former prosecutor turned law-enforcement agent. The two men instantly hit it off. They not only had the same philosophy when approaching a case, but both were willing to take risks and fly by the seat of their pants. They had a few other things in common as well. Each has a black belt in tae kwon do, and like Newcomer, Johns is obsessed with being the good guy wearing the white hat. They agreed to discuss strategy once a week. The game was afoot, and Kojima was their prey.

The Skype Connection

The highest enjoyment of timelessness . . . is when I stand among rare butterflies. . . . This is ecstasy, and behind the ecstasy is something else which I cannot explain.

—Vladimir Nabokov

This time I'm staying in touch *with the guy,* Newcomer promised himself. He sent Kojima an e-mail soon after their last meeting.

It was great to see you. I really enjoyed our time together. You're a good friend.

Then he promptly switched personas and dove back into Operation High Roller. That immediately threw a monkey wrench into things. He wasn't always around when Kojima tried to call, and the time difference made the situation difficult. Newcomer quickly learned that handling two undercover cases at once wasn't going to be so easy. When they did finally manage to connect, Kojima had one burning question. Had Ted Nelson signed up for Skype?

Damn! He'd been so busy with the roller-pigeon crowd that he hadn't had a chance.

"I'll get one of those little cameras and a microphone tomorrow and set it up," Newcomer promised. "Then we'll be able to see each other when we talk." *And I'll be able to get even more evidence.*

Kojima reminded him that it was also a good time to collect *indra* larvae. The cocoons that Kojima had brought back were already beginning to hatch. He'd send pictures so that Ted would know exactly what to look for. Then either Kojima or his son would come pick them up.

"Is Ken staying with his grandfather?" Newcomer asked.

Who knew? Maybe this really was a family-run business. Kojima's son helped to smuggle, and Grandpa cashed the U.S. checks.

"Yes, they're in the mountains. My father-in-law has a house there, also."

Kojima obviously had a very wealthy in-law. They agreed to speak by Skype the next evening.

"Okay. Don't play too much with young lady," Kojima said, signing off.

"You can never do too much of that," Newcomer replied with a laugh, and hung up.

Newcomer spent the next day buying a webcam, opening a covert Skype account, and surreptitiously aiming a second undercover camera at the computer screen in his home office. Then it was time to arrange exactly what Kojima would see.

Newcomer's home office tells more about the man than he'd like others to know. Outwardly reserved, he's clearly sentimental. The room is an ode to his childhood.

Old Dr. Seuss books tidily line one shelf. Arranged next to them are a series of toys including a plastic zookeeper, the Pink Panther, and Donald Duck, among other cartoon characters. A beloved stuffed bear keeps watch in a nearby corner.

A collection of Matchbox vehicles marks the next phase in Newcomer's life. There are models of his first Jeep, an *Adam-12* police car, and a Greyhound bus, the company his father once worked for.

His G.I. Joe collection stands rigidly at attention in a glass showcase. One doll wears a tae kwon do uniform that Newcomer had taken the time to sew. Their original boxes, all in tiptop shape, stand neatly beside them. A photo of three-year-old Ed with his dad hangs just above his desk.

The screen saver springs to life whenever his computer slips into sleep mode. Languidly lying on her side is a black-and-white cat with a paw seductively stretched toward the viewer.

Zuke is the pet Newcomer adopted while in law school. She lived with him through the worst times in his life—being broke, working as a lawyer, hating his job, going through multiple girlfriends, and too many moves. She never made him feel guilty and never got mad. Or, if she did, she simply bit him and moved on. Unlike his girlfriends, Zuke always forgave him and never held a grudge. There weren't any games between them, and she didn't pull crap. She was the perfect companion and Newcomer's closest buddy, the one creature that he felt he could trust. The word "no" meant nothing to her. The cat was rowdy and a scamp.

Zuke died of cancer shortly after Kojima turned Ted Nelson into Fish and Game. Her death helped to compound just how badly things were going at that time. It was the saddest moment that he'd experienced so far.

Newcomer carefully arranged the mementos so that nothing of himself would be revealed to Kojima. Then he methodically checked them once more. It was finally time for their first online video date.

He anxiously clicked on Kojima's Skype name, and Kojima promptly answered the line. It seemed almost unbelievable. He could actually look inside Kojima's house in Kyoto.

Kojima sat in a room filled with butterfly-rearing equipment, antique vases, and Japanese screens. The rumpled sheets on a futon bed were just behind him. The room looked as if it hadn't been picked up or cleaned in a year. In that sense, it was exactly the same as the apartment Newcomer had seen in L.A. In fact, so was Kojima. The guy looked as if he'd just rolled out of his unmade bed.

Kojima got right down to business. He discussed the ten pair of *chimaera* butterflies that Ted Nelson had ordered.

"My son must have put them away somewhere, but don't worry, I'll find them. I'm looking, looking," he said, and brought his face close to the camera as if to get a better glimpse of Nelson. He had on a pair of big round glasses that made him look like Mr. Magoo. "So, how do you like using Skype?"

"It's great. I like this a lot," Newcomer enthused. "It's nine PM here. What time is it in Japan right now?"

It was one in the afternoon. "But I'm always at home. I'm watching dirty movies all the time," he confided.

"Even during the day?" Newcomer asked with a laugh, unsure if Kojima was serious.

"Uh-huh," Kojima acknowledged. "All secrets we can discuss this way."

"That's right. Just you and me," Newcomer agreed conspiratorially, feeling a bit like James Bond.

"I do it for a lot of people. You can keep for your files. Prices and everything," Kojima replied, almost as if he knew what Newcomer was secretly doing.

"It's perfect," Newcomer concurred. "So, do you talk to your L.A. actor friend this way?"

"Yes. He can show me his big one on Skype." Kojima thought he saw Newcomer's lips twitch. "No, no. I'm just joking." Then he took a closer look at Newcomer. "Oh, you didn't shave yet."

Kojima still couldn't get used to the new version of Ted Nelson. There seemed something different about him this time.

Newcomer's fingers self-consciously wandered up to his mustache and lightly stroked it. "No, not yet. You don't like it, do you?"

Kojima gave a small shrug. "It's okay. You're nice-looking always. By the way, you can play for your girlfriend together also this way."

"Uh-huh," Newcomer responded, his mind focusing three chess moves ahead.

"Some people have a problem because they're showing each other," Kojima continued.

"Doing what?" Newcomer asked, having suddenly snapped back to attention.

"They play each other inside Skype together," Kojima tried to explain, but Nelson still didn't seem to get it. "That's okay. Your girlfriend is no problem."

What the hell is Yoshi talking about, anyway? Newcomer wondered, and brought the subject back to business. Nelson had spoken to his customer, who was interested in buying an *alexandrae* and a *homerus*.

Kojima abruptly cut him off. His friend had none in stock at the moment. "Besides, you need to be very careful when it comes to those two butterflies," he warned.

"You're right," Newcomer agreed. "I'm actually not going to sell them to anybody but this one person. He's very excited about it, and I trust him. So let me know when you can get them."

"I have twenty people waiting for them right now," Kojima advised. However, there were other butterflies that Kojima could easily get for Nelson, all of which were legal.

Newcomer did his best to steer Kojima back in the right direction. "Most of my customers really want birdwing butterflies. Do you have any of those that are interesting?"

Come on, Yoshi. I know you must have plenty of Appendix II butterflies stored in your house.

Kojima aimed to please. He promptly sent Newcomer a number of digital images via Skype. Newcomer clicked on file after file, uncovering a series of butterflies, each more beautiful than the next.

"This one is very difficult to get," he confided.

The creature was the size of a small bird, its wings a mesmerizing cinnabar with gold.

"I see it's called *Ornithoptera croesus wallace*," Newcomer noted, aware that it was an Appendix II butterfly. It was time to begin shopping. "I'd love to get one of those."

"But no CITES permit, okay?" Kojima said.

That's the idea, Newcomer thought grimly. Kojima would mark it as something else so that they wouldn't be caught. "I don't care about permits," Newcomer stated, spelling out the illegal act for his recording.

"Neither do I," Kojima concurred.

"I trust you, Yoshi. So, I'm not worried," Newcomer said, beginning to slowly reel him in. "And you trust me, right?"

"Uh-huh," Kojima agreed absentmindedly.

That was enough for tonight. Newcomer had just made an Appendix II purchase and couldn't have been more pleased. "Hey, you know what? I haven't eaten dinner yet, and I'm starving. I'm going to grab something to eat."

"Good. Then I can call you later," Kojima said. He was glad they were using Skype. Kojima got lonely and it was fun talking with Ted.

Newcomer glanced at his watch. It was already late, and he needed to get some sleep. He scrambled for an excuse. "Actually, I'm going to head over to my girlfriend's house after that."

"Oh, you can do anything you want. I can't stop you," Kojima replied, though he seemed to feel slightly insulted. "You are young boy still."

That didn't sound good. Newcomer knew he had to be available for Kojima no matter what. He quickly reviewed his schedule. *Damn!* He had to do a covert run on some of the pigeon guys tomorrow night.

"I'll be around this weekend and next week, too," Newcomer said, hoping to appease him. "Also, thanks for telling me about Skype. This is pretty cool."

"Really? If my friend from L.A. calls sometime, I'll call you, too. You want to see him, right?" Kojima coyly inquired.

"Sure," Newcomer responded. Who wouldn't want to talk to a TV star?

"Maybe he'll want to show you something," Kojima suggested with a giggle.

Whoa, this is starting to get weird. Is Yoshi serious, or is it just a joke?

"No, thanks." Newcomer brushed it off with a laugh. Maybe Kojima was just a goofball.

But Kojima was on a roll. "If I find a nice photo, I can send it to you. You can send me one, too."

Perhaps this was Kojima's version of becoming Skype pals. On the other hand, just what kind of photo was Kojima suggesting?

"Sure, you can send me whatever you want," Newcomer tactfully replied, and said good-bye.

He turned to see his wife, gave her a smile, and officially signed off. His work was done for the night. He could finally spend some time with her.

They'd just settled on the couch when his undercover cell phone rang unexpectedly. *Now what?* Newcomer thought as he saw that the call was from Japan.

"This is Yoshi. Call me on Skype. There's something I want to show you," Kojima instructed, and hung up.

Who knew? Maybe it was a Queen Alexandra. Newcomer had no choice but to do as he was told and log back on. Kojima appeared with a cup of tea by his side.

"Look, this is a Victoria. Do you see?" he asked, and held the birdwing butterfly up to the webcam.

Newcomer's heart beat a little faster. The butterfly's luminous green wings nearly shimmered straight through the screen. No wonder collectors got hooked on them. It was as regal as the British queen it had been named after. Newcomer covertly snapped a picture of Kojima holding the Appendix II butterfly.

"I have ten pairs of them. I can send you this one right away. Can you give me your address?" Kojima asked, morphing into the perfect salesman.

"Sure," Newcomer said, and then realized Kojima had yet to tell him what the price was. Did he deal with all his customers this way?

First bedazzle them and then hit them up for cash? "How much money do you want for it?"

Kojima would check with his son and get back to him.

Newcomer snapped another photo, imagining that the butterfly must have been even more beautiful when it was alive.

"This one is a very nice color," Kojima said, almost as if he could read Newcomer's thoughts. Then he suddenly began to cough.

"Are you okay?" Newcomer asked. For all he knew Kojima was still having heart problems.

"I'm okay. It's only the throat. I need . . ." Kojima mumbled something and giggled.

"You need what?" Newcomer prompted. He didn't want Kojima dropping dead in the middle of his investigation. Not after all the work he'd put into this case.

"No, I'm just joking," Kojima replied, and looked momentarily embarrassed.

It was tough enough dealing with Kojima's accent, but the cough made him impossible to understand.

"Anyway, just send your picture to me," Kojima reminded him, and said good night.

Kojima was turning out to be even more of a character than Newcomer had imagined, but nothing he couldn't deal with. All he had to do was stick to his plan of buying butterflies and somehow manage to keep Kojima happy.

Newcomer sent him an e-mail the very next day. He'd enjoyed their conversation the night before and looked forward to doing it again soon. *By the way, what's the price for that Victoria? I'll wire the money as soon I find out.* Newcomer was anxious to get the process rolling.

Two days later he morphed back into Ted Nelson, blue-collar worker, attending a pigeon-club fly on the weekend. The role was fun and easy to slip into. He was always accepted and liked by most of the men. He could schmooze with the best of them.

The fly was held at four different residences, including those of members Juan Navarro and Keith London. Newcomer took satisfaction in knowing that he'd already amassed evidence against both of them. The third fly was at the home of Eddie Scott. His place had two well-built sets of pigeon lofts, along with a two-story apartment residence on the back of his property. It was always interesting to see how these guys managed to hide what they did.

Scott regaled the group with stories of how he routinely stood on the second-story landing and shot falcons as they attempted to hunt his birds. It was a great spot from which to kill them. He even confided to Newcomer that the two-story residence had been built without the proper permits. If the city ever found out, they'd make him tear it down. Scott clearly knew what he was talking about. He was employed by the city of Los Angeles. Newcomer conscientiously made note of it.

Five days had passed since Newcomer last spoke to Kojima, and there was still no word about the price for the *Ornithoptera victoria*. *Man, you'd think this guy would have gotten back to me by now. Isn't this how he makes his living?* He tried not to worry, but one never knew with Kojima. The guy could turn on a dime, depending on his mood.

He dashed off a quick e-mail.

> *Hey, Yoshi. How are you? Is everything okay? I haven't heard from you in a while.*

Kojima immediately responded.

> *Hi, Ted. I haven't been able to call since you never turn your Skype on. I'm leaving for the Philippines with my son tomorrow and don't have time to deal with it right now. I'll be in touch when I get back.*

Damn! Newcomer could kick himself. This was one reason why Kojima had dropped him the first time, and it was the last thing he needed to have happen again. The slightest mishap could screw up the case. He didn't wait but shot back an e-mail and profusely apologized. *Sorry. I thought once Skype is set up that it automatically comes on. I'll take a look and figure out why you had trouble.* He was determined to make it work even if he had to take the damn thing apart cyber piece by cyber piece. *Anyway, have fun in the Philippines, and let me know the price for the Ornithoptera victoriae when you return.*

It was like trying to stick his finger in a dike and praying another problem didn't pop up somewhere else.

Newcomer's workload now became heavier as he juggled cases, doing paperwork in the office and garbage runs and surveillance at night. Saturdays and Sundays were always spent at roller-pigeon flies. As a result, things at home began to get even rougher. Allison's summer wasn't as busy as the rest of the year, and she wanted to spend more time with him. The situation grew increasingly tense as Newcomer was drawn deeper into both cases. To make matters worse, Allison's dog of fourteen years wasn't well and Newcomer became distant as he buried himself in his work.

Newcomer spoke with Kojima by Skype two days later. Kojima appeared on-screen wearing a rumpled green shirt and navy shorts, with a fanny pack around his waist. Oversized glasses encircled his eyes, while his hair stuck out randomly. Once again, it looked as though he'd tumbled out of bed, although it was evening in Kyoto. He spoke as if everything was perfectly fine. In fact, things couldn't be going better.

He'd just traded a pair of CITES Appendix II butterflies for one hundred *indra* pupae from an American collector. The two butterflies that he'd sent would normally go for a total of two hundred dollars. By comparison, Kojima would be able to sell each pair of *indra* for at least that amount and make a tidy profit.

"The guy contacted me on the Internet. He has *kaibabensis* and *martini indra.* I don't know how he can get them. The *martini* is one hundred percent from inside Grand Canyon National Park," Kojima boasted, clearly thrilled at having gotten the better end of the deal.

The United States protects both *P. indra* species. The pupae wouldn't be sent to Japan. Instead, they'd go to his father-in-law's house, where Kojima would pick them up.

"Now I want to show you some beautiful butterflies," Kojima said, and pulled open a large book. Turning the pages, he pointed to one. "Do you know what this is?" he asked with a sly smile.

If this is a test, I'm going to fail it, Newcomer mused.

"Hold on. Just lower the book a little," Newcomer instructed, having no clue as to what butterfly he was looking at. "You mean that little yellow-and-black one?"

It wasn't a big flashy birdwing, so why was Kojima bothering to show it to him?

"This is CITES I," Kojima said

Whoa! Hold it a second. Newcomer took a closer look. "Exactly which butterfly is this?" he asked, his interest having been sparked.

"This is a *hospiton* from Sicily. You know Italy? Al Capone? Everybody wants it. But it's on the endangered species list, so no one can get," Kojima said, his finger lingering lovingly on the page.

Was this some sort of a tease, or was Kojima actually providing an opening? Newcomer paused a moment, then decided to go for it. "That's very impressive. I wouldn't even know how to go about getting one of those."

Kojima didn't miss a beat. "If you have a customer, I can sell to you. This one we just write down *machaon.* They look very similar, only the *machaon* is a legal butterfly. No one can tell the difference between the two."

Newcomer felt as if he'd just downed a six-pack of Red Bull. His pulse raced as adrenaline shot through him. He couldn't believe

Kojima had just offered to sell him a CITES I butterfly. *Play it cool,* he reminded himself.

"I'll ask one or two of my customers and see if they're interested. How many of those can you sell? Just a few?" he fished, hoping he sounded a lot calmer than he felt.

"I can give to you for seven hundred dollars. I used to have twenty pairs of them, but almost everything is gone. I only have three pairs of *hospiton* left."

"Holy cow!" Newcomer exclaimed, unable to stop himself. *How did Kojima obtain twenty pairs of a butterfly deemed nearly impossible to find?* "And you have ten pairs of the Victoria?" he double-checked.

"Yes, those are eighty dollars a pair," Kojima informed him.

Newcomer quickly added it up. He'd order two pair of Victoria to start and see if they passed through Customs. If not, they'd work something else out. "You just sent a package of CITES butterflies to an American collector, and he got his package, right?"

Kojima assured him there was never any problem. He sent packages to people in the States all the time.

"Okay. Then I want all ten pair of Victoria and I'll get back to you about buying the *hospiton.*" Newcomer was on a butterfly high, having never felt more exhilarated in his life.

"Eight hundred dollars is big money for ten pair of Victoria. Are you sure you're okay with that?" Kojima asked thoughtfully. Nelson was acting like a big-time dealer. Where was this coming from all of a sudden?

"Yeah, absolutely no problem. I just sold the boat business for a hundred fifty thousand. That gives me some play money to put toward my new career. Buying and selling butterflies is all I plan to do from now on. Besides, I trust you, Yoshi. I know you'll send the rest, assuming all goes well," Newcomer said, hoping to whet Kojima's appetite.

With that, the line went dead.

Oh shit! Newcomer thought. *Did I say something wrong? Does he know that I'm lying?*

He breathed a sigh of relief as the musical ring tone of Skype played again.

"I'm sorry, man. I think I touched the wrong place and accidentally hung up," Kojima apologized with a giggle.

For a big-time smuggler, Kojima was a hell of a wacky character. Newcomer agreed to speak to a couple of his rich customers about buying the *hospiton*.

"Okay, but it's best to tell them by phone and not in an e-mail," Kojima advised.

"For safety reasons, huh?" Newcomer queried. He was actually learning a lot from this guy.

"Yes, for CITES it's always better that way. Also, remember to only take cash."

He sneezed and explained that he had a cold.

"That's because you've been running around too much naked," Kojima joked.

"Yeah, that must be it. I haven't been feeling well lately. My body is sore," Newcomer replied.

"Maybe too much using for the hand," Kojima jested, and both men laughed. Kojima enjoyed Skyping with Ted and it helped to break up his day. "You really must take care of yourself and feel better. You have a cheap apartment somewhere?"

"Do you know where Long Beach is?" Newcomer replied unthinkingly.

"Long Beach? Yes, that's a very nice place," Kojima remarked.

"Yeah, I live . . . " Newcomer immediately caught himself. Damn it! He really must be sick. He'd almost told a smuggler where he and his wife could be found. *Why don't I also give him my real phone number and invite him over for dinner while I'm at it?* "No, I used to live there but I live in Torrance now."

He'd have to be more careful from now on. He deftly brought the subject back around to butterflies.

"People are very afraid of Fish and Wildlife here in the States. I've even heard that special agents are at the insect fair sometimes," he offered, curious as to Kojima's response.

"They're always at the show," Kojima acknowledged with an indifferent shrug. "Everybody says to me, 'Yoshi, they are coming to your table.'"

"They're probably watching you," Newcomer teased. Little did Kojima know that he was looking into the eyes of one right now.

"The L.A. show is mostly ninety-nine percent amateurs," Kojima scoffed. "The Tokyo fair is where all the real professional collectors go. The sellers have lots of CITES butterflies there, and no one has CITES permission."

"Wow, I wish it was that way over here," Newcomer said.

"You can come to my house and I'll take you there. I'll pay for the train. Everything. We'll stay at the number one hotel in Tokyo. You can also help me at my table," Kojima offered generously. That was his way. He liked to help out young collectors.

"That would be great," Newcomer said, beginning to drift along with the conversation.

"Of course, my boy might be here, so you must sleep with him," Kojima nonchalantly remarked.

Newcomer instantly jerked to attention. "What?" he sputtered as if gasping for air. Had he heard Kojima correctly? Or had there been a miscommunication?

"He has a bigger bed, so you can stay with him if he's here," Kojima patiently clarified.

For chrissakes, don't go jumping to crazy conclusions, Newcomer thought with a sigh of relief. Talk about "lost in translation." Sometimes he wasn't sure what the hell Kojima was saying.

"Sounds good. Listen, I'm going to eat and then rest for a while," Newcomer said, itching to end the conversation.

"I'll send a package to you Express Mail in the morning. Let me know when you receive it, and perhaps we can do more business together," Kojima suggested wistfully.

"Thanks, Yoshi. I appreciate it." Newcomer glanced at the clock. It was 9:50 AM in Los Angeles. That meant it had to be 1:50 in the morning Kyoto time.

"Go to sleep. See you soon. Yes?" Kojima asked softly, drawing the conversation out a few more precious moments.

"Yes. Good night, Yoshi. Sleep well."

Newcomer smiled at him and ended the call. It felt as though he'd just said good night to his girlfriend.

NEWCOMER SPENT THE DAY playing catch-up and filling Palladini in on both cases. He was continuing to get evidence of hawks being killed on surveillance tapes. Then he gave her the other piece of good news. Kojima had just offered to sell him three pairs of Appendix I *hospiton*.

"You're kidding! That's unbelievable," Palladini said, and congratulated him.

Newcomer didn't say so, but he thought it was pretty amazing, too. However, he still couldn't understand why Kojima had flip-flopped from being so cautious to suddenly seeming reckless. Evidently, he truly believed he'd never be caught and was now convinced that Ted had become his compatriot. Even so, some of his remarks were pretty odd. They'd caught Newcomer off guard, and that was hard to do. Maybe it was just Kojima's way of bonding with men. In any case, he didn't intend to speak with Kojima for the next few days. There were other matters that needed his attention.

Newcomer went home that night planning to spend a quiet evening with his wife. It might help heal the rift that was developing between them. He was caught by surprise when his undercover cell phone rang. The insistent jingling came from where it lay buried deep inside his briefcase. It ended abruptly as he finally pulled it out.

Wouldn't you know? The call had been from Kojima. What could he possibly want again so soon? They'd spoken only that morning. Damn, he'd have to check in and find out. Newcomer left Allison with her dog, Huck, and went into his office. Once again, he made sure that nothing personal was in sight and then logged on to Skype.

Kojima popped on-screen looking exactly as he had twelve hours ago. He must not ever bother to shower or change, simply wearing the same clothes all the time.

"Hey, Yoshi. Did you just try to call me?" Newcomer asked.

"Yes, but you never answer me," Kojima replied testily.

"You know what? My phone's in my briefcase and I knew it was you, so I just came to my computer," Newcomer explained.

"Fuck you, man," Kojima replied.

"What?" Newcomer asked, momentarily taken aback. Kojima had never cursed at him before. *Something must be bugging him.*

"I said 'Fuck you, man,'" Kojima repeated with a laugh.

"Well, that's not very nice. By the way, I'm feeling much better. Thanks for asking," Newcomer retorted as his landline erupted in a series of rings.

"Oh, is that your girlfriend calling you?" Kojima inquired snippily.

What an ass, Newcomer thought, and grabbed the line before his answering machine could pick up. A fellow agent was on the wire.

"Hey, I'm talking to my friend in Japan," Newcomer said, aware that Kojima was hanging on his every word. "Call me back in an hour."

"Your girlfriend?" Kojima coyly asked again, as Newcomer hung up.

"Yeah, that was her," Newcomer replied archly. He could almost swear Kojima was yanking his chain for having a personal life. Maybe he needed to get out from behind that computer of his more often.

Kojima had the tracking information for the package that was mailed that day. "I also added fifteen common butterflies. That way even if the box is opened and checked, the Victorias won't be easy to spot. They're mixed in with the decoys,"

"That's very clever. You're a smart guy, Yoshi. I like that," Newcomer said, making a point of flattering him whenever possible. "That reminds me, I talked to my customers today about the *hospiton*. They definitely want to buy them."

This was the moment of truth. Would Kojima actually make good on his offer? *Come on, Yoshi. Deal or no deal? It's your move.*

"Remember I told you about the *machaon*? The butterfly that looks similar? I'll mix both species together and mark them with a code. That way only you will know which are the real *hospiton*," Kojima said.

Newcomer was duly impressed.

Huck chose that moment to come padding into his office. The dog panted loudly enough to have been a huffing and puffing engine. Kojima's mind was obviously elsewhere, as he didn't seem to notice. Instead, he glanced away from the screen as if having heard an unexpected noise in his own house. Then he went back to fiddling with some butterflies on his desk.

"That's my boy. He returned home early this morning after being out all night. He's still sleeping. I want to fuck him," Kojima remarked diffidently.

Newcomer's stomach clenched tight as a ball during a moment of stunned silence. This time there could be no denying what the man had said. Newcomer tried to cover his discomfort with a nervous laugh. "What? Your son?" he asked. His distaste for Kojima had just taken a giant leap forward.

"Yeah," Kojima replied casually, and smoothly moved on to the next topic. "Do you like to drink?"

"A little," Newcomer replied tentatively, curious as to what might be coming next.

"Good. When you come to Japan I can give you some sake."

Uh-huh, and then what? Possibly a threesome with your son? Kojima had just placed himself in the definitely weird category.

Newcomer no longer had any lingering doubt as to Kojima's sexuality. Now he was worried about how twisted he might be.

Kojima suddenly looked off to the side of the room again.

Uh-oh, Newcomer thought, his stomach twisting once more.

"My boy is still sleeping. You can see him naked. Do you want to take a picture?"

What's going on? Is this his idea of some kind of test? Say the wrong thing and bzzzzz, *you're thrown out of the game? Just play along,* Newcomer decided.

"No, thanks," Newcomer replied.

"You know, I've been diabetic for almost thirty years now," Kojima offered casually, as if it was an excuse for his behavior. "I take insulin every day. It's because of diabetes that I had to have bypass surgery."

"Sorry to hear that," Newcomer commiserated, not sure of what else to say.

Kojima sighed deeply and shook his head. "It causes so many problems. I cannot see well. I cannot see your beautiful body, for instance."

Newcomer couldn't help but grin. "That's a good one, Yoshi. My camera only catches my face."

Kojima gazed at him straight through the webcam before lowering his eyes as if toward Newcomer's crotch. "I want to see your beautiful body and everything. Leave your computer on all day. It will be easier for me to catch you that way. Now tell me, what do your customers want?"

So this was the game— butterflies and Internet sex. That's why Kojima was becoming bolder and more brazen with each call. Now it was beginning to make sense. In that case, two could play.

"They're interested in *hospiton* and *homerus*," Newcomer said, boldly stating two of his demands.

"I already told you. I don't have *homerus* right now."

Maybe not, but Kojima certainly had lots of Appendix II specimens that were available. He bragged of selling three to four hundred dollars' worth of butterflies every day and had thirty steady customers in the United States.

"How do you know who you can trust when sending CITES material?" Newcomer asked, wanting to learn how Kojima continued to get away with it.

"I don't mind because I'm in Japan and they're in the U.S. I tell them if you have a problem, just refuse the package. Fish and Wild can't touch me, and Japan doesn't care. That's why I tell you to be careful of selling CITES butterflies in the U.S. It's much more difficult," Kojima warned. "CITES I, I don't like to sell unless my customers pay cash. Checks provide evidence."

Kojima was beginning to sound bored, so Newcomer decided to talk about something else for a moment. "I think I might shave off my mustache," he said, his fingers playing with it suggestively.

"You shave all your hair is better," Kojima said, instantly perking up.

"You mean you want me to shave my whole head?" Newcomer chuckled. Maybe Kojima liked bald guys.

"Yeah, everything, and then show me next time. Get on camera and show me everything," Kojima demanded.

"Yeah, sure," Newcomer said. That was enough of a diversion for now. It was time to get back to business.

"Hey, one of my customers said he wants to collect all three of the most endangered butterflies. The *chikae*, the *alexandrae*, and the *homerus*." *What the heck. The* hospiton *is already in the bag. Now I just have to get the rest.*

"The *chikae* comes from the Philippines. It's very difficult to get them out of the country, but I can sell you one from my own collection," Kojima offered amiably.

"I don't want to take it from you unless you don't mind selling it," Newcomer replied, somewhat surprised. Wouldn't he want to keep his own collection intact?

Kojima dispelled any notion of that. "I don't care as long as I can make money," he stated flatly.

So much for any personal attachment Kojima had for butterflies. He was in it for the moolah.

"That would be great, but my customer still wants all three species. What about the Alexandra? Any chance of that?" Newcomer pressed. Getting an Alexandra would be nearly as good as snagging the Hope Diamond.

"You sure your guy is okay?" Kojima asked. One could never be too careful when it came to CITES I butterflies. They were the ultimate prize, and an FWS agent might be trying to nab him.

Newcomer swore there was no problem. He'd sold CITES butterflies to this client before. The man wouldn't turn them in.

"Okay. If you say so," Kojima conceded.

Ted Nelson was turning into such a good customer that Kojima sent him a photo of another expensive butterfly, the *Papilio agehana maraho.* The female sold for a minimum of five thousand dollars. The butterfly was protected in Taiwan and found only in its national parks. No one could get them. Except for Kojima, of course. He received them every month. The butterfly had a broad, lobed tail and wings of black velvet. Each bore a large white square and was rimmed in red that was rich as blood.

"How are you able to do that?" The more Newcomer learned about Kojima, the more determined he became to unravel the puzzle. The man was like an onion whose layers didn't end.

"That's a secret," Kojima replied with a mysterious smile.

He sent Newcomer more butterfly pictures on Skype. There were photos of gorgeous red-and-black butterflies that Ken had recently brought back from Cuba. He had six hundred of one particular species that sold for $250 apiece and a couple thousand of another Cuban specimen. The numbers were mind-boggling. Kojima and his son were like two John Deere harvesters reaping every butterfly in sight.

"How did your son manage to get so many butterflies out of Cuba?" Newcomer asked.

"We have permission from National Geographic because we work for them," Kojima replied.

Of course. What better way to pull legal strings?

"Wait, I'm looking for Ken's photo for you," Kojima told him.

"You already showed me on your cell phone. Remember?" Newcomer said.

"This one is better. It's all naked. He's very exciting," Kojima said, barely able to contain himself. "I can't find it right now, but it will make you crazy when I do."

"Yeah, *you're* making me crazy," Newcomer said, laughing ruefully and continuing to play along.

Instead, Kojima now plied him with photos of *Troides aeacus*, *Ornithoptera paradisea*, and *Bhutanitis lidderdali*. Also known as the Bhutan glory, the butterfly touted spiky tails and an arresting display of scarlet, yellow, and black mosaics on a backdrop of brownish wings. Newcomer's head began to spin, the last file seeming to take forever to download.

"Boy, this must be a big one," he commented, beginning to tire.

He couldn't have given Kojima a better setup line if he'd tried. "Big one? Oh, you like the big ones!" Kojima crowed in delight.

The guy is relentless! Think of it as a game of tennis, and just keep hitting them back to him, Newcomer told himself. "I think *you* like the big ones. So, how much do you make a month selling butterflies, anyway?" He might as well get some information in between all the double entendres.

"Depends, but is about thirty to fifty thousand dollars a month," Kojima boasted.

"Wow! That's pretty good, Yoshi." *You'd think he could at least afford a housekeeper, bringing in that kind of money.*

The amount was impressive, but the very thought of it sickened him. That translated into thousands of butterflies being removed from the wild. For some, it was a step closer to extinction.

"Sleep well. Don't do anything tonight," Kojima teased, and said good night.

Newcomer turned off Skype and sat for a long time gazing at Zuke's image on the computer screen. Kojima had now offered to sell him two different Appendix I butterflies, the *Papilio hospiton* and the *Papilio chikae*. He knew it was the first step in a complicated chess game to catch the man. Especially since Kojima had made it clear there were certain things that he expected.

The situation had taken a bizarre twist. Newcomer was no longer just dealing with butterflies, he was dealing with a man focused on him as the target of his desire. This was a game that he'd never before played, and he had a choice to make. He could get out now or stick with his plan and do whatever was necessary.

Newcomer knew this would be a true test of his undercover skills. It would also prove to be the role of Newcomer's life.

Mind Games

Obsession fulfills passion's choice of destiny.
—Daniel Waldschmidt

Ted Nelson received an e-mail from Kojima the very next day. Contained in it was the address of Kojima's bank, along with his account number and all other pertinent information.

Terrific, Newcomer thought. Their business together was moving forward. He could wire money to Kojima and, in return, receive packages of smuggled butterflies. Everything was in place.

His next move was to call the National Geographic Society.

Someone there either knows what's going on or they're in for a hell of a shock about one of their employees, he mused. It was time they knew that a smuggler was in their midst. Kojima claimed to have worked for them for years and continued to brazenly use their name. Saying the words "National Geographic" was akin to uttering the phrase "open sesame." It opened important doors and allowed Kojima to obtain beetles and butterflies that were otherwise illegal for export. Kojima had been milking the association for all it was worth.

Newcomer's call was directed to its Department of Human Resources, and he wondered what they would say. Just how far up the line did Kojima's shenanigans go? Would it possibly be high enough to smear the society's good name and bring on a world of pain and trouble? He was about to find out.

Special Agent Newcomer inquired if National Geographic had one Hisayoshi Kojima on its payroll either as an employee or a subcontractor. The voice on the other end said that the matter would be looked into and someone would get back to him as soon as they knew. Newcomer decided to give them a couple of weeks and call again if they hadn't responded by then.

A flurry of e-mails between Newcomer and Kojima now ensued over the next few days. Business heated up for the sale of Appendix I and II butterflies. Newcomer did his best to play it cool.

Ted Nelson was still waiting to hear from one customer but felt sure he'd have an initial order of at least seven hundred dollars. Another client first wanted to see the sample of *Ornithoptera victoriae* that was coming. Based on that, he'd probably order another three or four pair. Then there was a customer chomping at the bit to buy the Appendix I *hospiton*. A fourth client was also interested in purchasing the illegal Appendix I butterfly. However, he was only willing to pay half of the money up front.

I think he doesn't believe I can really provide them—ha ha. I might place one order next week and another soon after, Newcomer wrote to Kojima.

Could Ted possibly Skype with Kojima that evening? He was driving to San Diego for a tae kwon do tournament that night but would Skype Kojima at around 9 PM from his hotel.

Newcomer figured that should hold Kojima for a couple of hours. *As long as I always have that lifeline open, then the case will be safe,* he told himself. He wouldn't allow for any mistakes to happen this time.

Newcomer drove to San Diego after having checked and double-checked that the hotel had high-speed Internet. He arrived only to learn that someone on the hotel staff had screwed up. High-speed Internet wasn't available.

Damn! Newcomer broke into a cold sweat, his reaction as reflexive as one of Pavlov's dogs. He already knew what would happen, and he proceeded to fly into a panic. He was supposed to Skype with Kojima at nine, and it was already later than that.

Let me check my cell phone. Maybe Yoshi left a message on there for me. Newcomer hit it on the mark. Kojima had already left several.

Where are you? Why aren't you calling me?

Each message was more desperate and demanding than the last. Oh God, it was déjà vu. Worst of all, though his pay-as-you-go cell phone would accept international calls, it wouldn't allow Newcomer to make them. The only way he could contact Kojima was by e-mail, and it was impossible to do that at his hotel. This was a Mylanta moment if ever there was one as history began to repeat itself.

It was now after midnight, and Newcomer was growing increasingly frantic. He had to find a way to contact Kojima, or the case would implode. He finally stumbled upon an Internet café that was still open. He promptly rented a terminal and logged on to his undercover e-mail account, where even more messages from Kojima awaited him.

I can't get hold of you. I need to talk to you!

The words virtually jumped off the screen and pummeled him. Newcomer hadn't recognized Kojima's pattern in 2003 until it was too late. Now he knew exactly what was happening. Kojima would pull the plug, and Newcomer would be back to square one. He was so close to getting butterflies that he couldn't believe it might possibly blow up in his face for a third time. At this point, he'd do whatever it took to staunch the flow.

Newcomer pounded out an e-mail that was even more agitated than those left by Kojima.

I am at a TERRIBLE hotel!!! I will never stay here again. They promised me high-speed Internet, but there is only a phone line. I'm so mad I am screaming.

I am going to watch a Taekwon-Do tournament tomorrow in San Diego. Then I'm going back to LA. I will Skype with you tomorrow night at about 8 or 9 pm LA time.
Is that ok? Sorry, Yoshi. I'm so mad you can't believe it!!!!!!!

Kojima reached him by cell phone a few minutes later.

"Yoshi! Did you get my e-mail?" he asked nervously.

"Yeah, that's why I'm calling. I can't talk to you tonight," Kojima replied nonchalantly.

What! Then why leave all the damn phone messages and frenzied e-mails? Son of a bitch! The guy was jerking his strings and playing him again!

Newcomer knew it was all part of the game. Kojima was a master manipulator who was now turning the screws. This was his punishment for having kept Kojima waiting. Newcomer also knew what was expected of him—a good deal of apologizing and begging.

"I'm so sorry, Yoshi. There's no way I can get on Skype at this hotel. It's a slow connection," he began to explain.

"Oh, I see," Kojima replied with the slightest tinge of petulance. If Ted couldn't keep his word, then he shouldn't give it in the first place.

"I'm so mad at them, I can't tell you." Newcomer did his best to appease, but Kojima held all the cards and Ted Nelson was clearly being taught a lesson.

"It's okay." Kojima finally relented, satisfied that Newcomer was back in line and he once again had control. "Did you see my Web site with its announcement of all the new butterflies for sale?"

Newcomer's heart beat normally again for the first time in hours. "Yes, it looks great. Has anyone else offered to buy the *hospitons* from you?" *Please don't sell those out from under me.*

"Don't worry. I hold them for you right now," Kojima assured him.

Newcomer had a limited amount of buy money from Fish and Wildlife and needed to make it last as long as possible. He'd have to come up with a plan. It was time for Nelson to launch into his problem.

"One of my customers only wants to pay three fifty for a pair of *hospiton* up front and the other half when he actually sees it." Newcomer was so tired that he wasn't thinking properly and was unprepared for Kojima's response.

"And you can make a profit at that price?" Kojima asked cynically.

Newcomer was momentarily speechless and then quickly realized his mistake. He was buying the pair from Kojima for seven hundred and so couldn't possibly resell them for the same amount.

"Sorry, I just got confused for a moment," Newcomer said, trying to tap-dance his way out of it. "Actually, I'm going to charge him twelve hundred for the pair."

Kojima grunted in dissatisfaction. "You do better than me."

Newcomer's head began to pound. There was absolutely no winning with this guy.

"So I guess he's going to pay six hundred up front and then another six hundred after he sees them. He's skeptical that I'll be able to get *hospitons.* That's why he doesn't want to pay cash for them all at once," Newcomer continued, hoping that Kojima would bite.

"If he doesn't want to pay cash, then take a check," Kojima replied.

The suggestion hit Newcomer with the power of a sucker punch. *Wait a minute. Isn't that precisely what Kojima has been telling me not to do all along? So why the sudden change of heart?*

"Yeah, but I thought checks create an evidence trail and that I want to be careful," Newcomer retorted.

"You're too careful," Kojima replied lackadaisically. If Ted Nelson wanted CITES I butterflies, then he'd have to be willing to pay for them. Kojima wasn't running a charity organization.

"Hold on. You taught me that you can't be too careful," Newcomer shot back. *What game is Yoshi playing now?* There were so many of them that it was hard to keep up.

"Somebody catch you, then they're going to catch me. Same thing," Kojima rejoined slyly.

"Except that you're in Japan, so nobody can get you," Newcomer countered.

It had been a long, nerve-racking evening, and Newcomer was totally fried. Kojima obviously didn't care what happened to him as long as he got his money. Newcomer was about to respond when Kojima's shop buzzer rang, jolting his nerves like a powerful electrical current.

"I'll talk to you tomorrow," Kojima said, and abruptly hung up, having clearly won this round.

NEWCOMER HEADED BACK TO LOS ANGELES the next evening. However, he chose to see a movie with his wife rather than Skype with Kojima. They hadn't spent much time together, and he needed a short break from the man. He sent Kojima an e-mail later that night. He'd taken a look at Kojima's Web site and wanted to purchase two additional butterflies.

> *How much for* O. croesus wallace *and how many do you have available? Also, how many* Troides helena *and how much?*

Ever the conscientious businessman, Kojima promptly responded. It was too bad, but Newcomer was out of luck. He had just sold the last of those butterflies. Nelson would have to learn to be quicker the next time around.

Newcomer sighed deeply. It was yet another lesson that the case could explode on Kojima's whim. He had no choice but to toe the line.

He put on his mental suit of armor and Skyped Kojima the following night. Once again, everything seemed fine. But then Kojima was in an unusually good mood. His Las Vegas friend planned to collect more *P. indra kaibabensis* from the Grand Canyon the next day and send Kojima another hundred pupae. That would mount up to an additional twenty-five grand in Kojima's pocket.

"Maybe your friend will give me some pointers so that I can collect for you, too," Newcomer suggested. He couldn't think of an easier way to take down one of Kojima's suppliers. All he needed was a little cooperation.

"No, I don't think he will, but I'll ask," Kojima replied. He wiggled in his seat as if some of his insects had hatched and worked their way inside his pants as he continued to check his online sales, blow his nose, bite his fingernails, and loudly slurp his tea. Just watching him made Newcomer feel fidgety.

"So how many CITES II butterflies do you have in your house right now?" Newcomer inquired, as Kojima rang up another sale from his Web site.

"Maybe a thousand," he revealed.

"Really? You keep that much in stock?" Newcomer asked, all the while recording the information. This guy was the Costco of butterfly dealers.

"Yeah, but those sell out in just one month's time. Then I get more."

"Wow. How much money do you think you'd make if you sold all your CITES butterflies tomorrow?" Newcomer asked, ticking off his list of questions one by one.

"Maybe a half million dollars," Kojima said.

Forget Costco. Kojima was more like Fort Knox. He even claimed to have a butterfly that was a cross between a *victoriae* and an *Ornithoptera priamus* called *Ornithoptera allotei,* which was Appendix II, very rare and extremely valuable. The price for one pair was thirty thousand dollars. Newcomer had no idea if that was true or if Kojima was simply blowing smoke up his pants.

"Oh yeah? How many of those do you have?" he asked, trying to check the information on Google.

"That's a secret," Kojima said, adding one more mystery to Newcomer's burgeoning list to be solved. "Crazy Japanese customers buy them. Sometimes a crazy American will too."

"So if I come up with thirty thousand, you'll sell one to me?" Newcomer asked.

"Of course. If you show me your beautiful one, I'll even give you a discount," Kojima offered.

"What? Like $29,500?" Newcomer teased.

"More like $299.99," Kojima retorted.

"That's all I'm worth to you?" he quipped. He would have thought he could bring in more than that. Newcomer studied his image on the bottom left-hand corner of the screen and carefully smoothed his Fu Manchu mustache. The money for all the butterflies he'd ordered so far was beginning to add up. He figured he might as well look good for Kojima and see if something could be worked out. "Listen, I might have to make two bank transfers to you, Yoshi. I'm making my customers pay in advance, but I'm still waiting for some of the money."

"That's all right. Just send it to me all at once," Kojima replied cordially.

This was something new. Maybe Kojima was beginning to loosen up. "Really? You don't mind waiting a few days?" Newcomer double-checked.

"If you can show me your beautiful body, it's okay," Kojima threw out.

Of course there had to be a catch, and everything always came back to sex. So, which was Kojima was more obsessed with—seeing Newcomer's body or selling him butterflies?

Newcomer genially laughed. "No, I don't do that for money. That would make me a whore, wouldn't it?"

"That's okay. I'm just joking," Kojima said, and swiftly backed off. There was no reason to scare the boy.

"Hey, I may have to go to Denver for a few days to help my mom," Newcomer said, tired of the game and wanting to change the subject.

"Oh, that's nice. How old is Mommy?" Kojima asked, seeming genuinely interested.

"She's seventy-five. How about your mom? Does she live with you?" Newcomer asked.

For a moment, Newcomer felt as if they were simply two friends talking.

"No, she's smarter than that. I'm crazy." Kojima laughed. "One day I came home and she's not here anymore. She lives with my brother now."

Kojima almost looked sad as he lifted his leg onto the bed and slapped it. "My leg is no good. Everything's no good. Not even the center. Is your leg okay?" he asked.

It seemed an odd question. "Yeah, but nobody took a vein out of mine," Newcomer reminded him.

"No, I'm asking about your center leg," Kojima carefully enunciated.

I must be getting used to him. Half this stuff isn't bothering me anymore. "Oh, yeah. The third leg is always good," he jested.

"Oh really? Next time you show me," Kojima suggested with a giggle.

"I don't think my camera's big enough to catch it," Newcomer retorted.

"Oh, I can catch," Kojima assured him.

Newcomer suddenly turned serious. "Hey, Yoshi. Do you *really* have a half million dollars' worth of CITES butterflies in your house?"

Kojima looked at him and slowly nodded. Their eyes locked, and for the briefest moment, it almost seemed as if Kojima knew what might be awaiting him.

🦋 🦋 🦋

NEWCOMER BELIEVED HIS WORK to be done for the night as he sat in front of the TV with Allison. There was no such luck. The familiar ring tone of Skype sounded in his office less than two hours later. Kojima was calling again. Newcomer trudged in and answered to see Kojima's image pop up on the screen.

"I just spoke to my Las Vegas friend. I think someone probably turned him in for illegally collecting *indra kaibabensis* pupae once. He said he's too nervous to show you," Kojima informed him. "By the way, he says you're a Fish and Wild guy."

It was as if the floor fell out from beneath Newcomer. *Does Kojima's friend possibly know who I am?* Newcomer wondered, doing his best to keep his poker face in place. "I'm a wild guy, but I'm not a Fish and Wildlife guy," he said calmly.

"Then show me," Kojima challenged. "I put on my glasses and you can show me your beautiful one."

Shit. Yoshi's too damn smart. He knows perfectly well that a federal agent could never do such a thing.

Newcomer deflected the dare with a laugh. "That's a bunch of BS. You're not going to trick me that easily. Hey, I have to eat dinner now. I'm going to get off."

"Go ahead. But don't eat too much the big long john," Kojima said, chuckling.

This guy never gives up. "No, just enough," Newcomer said, playing along, and ended the call. That had been way too close for comfort.

As it turned out, the night was still young. Newcomer's Skype line rang again a short while later. He was attached to an invisible leash being jerked. Allison simply gave up. She took the dog and went to bed. Newcomer barely even noticed.

"Did I wake you up?" Kojima asked as Newcomer appeared on the screen.

"No, I'm just watching TV," he replied.

Kojima gave him a list of four more Appendix II butterflies that were now available for sale. They ran the gamut in price from fifteen to a thousand dollars per pair. "I also have two more friends that are going to collect in the Grand Canyon for me. I plan to come to the U.S. in July and pick up the pupae," Kojima informed him.

"Oh, yeah?" Newcomer said, perking right up. "Let me know when you're coming and you can stay with me if you need to." He'd have a nice jail cell ready and waiting with Kojima's name on it.

Kojima giggled like a teenage girl at the suggestion. "Oh, I'm so scared!"

"So who are these two guys anyway?" Newcomer asked, hoping for more information.

"Just some collectors. But my friend Thomas is really wanted by the Fish and Wild guy for illegally collecting butterflies in all the national parks," Kojima offered, as if dangling bait. "Did you send me your beautiful picture yet?"

"No, not yet. I'll send you something soon," Newcomer promised, pretending to abide by the rules.

"I want to see everything," Kojima commanded.

Newcomer dodged it with a laugh. "It's eleven-thirty here. It's time for bed. I have to get up early tomorrow morning."

"You going to bed alone or with your girlfriend tonight?" Kojima asked wistfully.

"No girlfriend tonight," Newcomer firmly replied.

"In that case, I'll send my boy."

Why does Yoshi keep trying to push his son on me? Newcomer wondered. "That's a long trip."

"It's okay. He looks like an international hooker. We'll charge your credit card," he said, guffawing.

Newcomer listened in amazement. It was a hell of a way for Kojima to talk about his own son.

"I'm terrible," Kojima said, almost as if he had read Newcomer's mind. "Listen, after tomorrow I'm going to send you *hospiton.*"

Any drowsiness that Newcomer felt was instantly dispelled.

"You can go on InsectNet's Web site and download labels to print whatever information you want. It looks more professional that way. Just don't fill in your name as the collector," Kojima advised.

Newcomer quickly typed in InsectNet's URL on his computer, and the Web site popped up. *Son of a gun. Yoshi is right. Look at that. Insect labels can easily be downloaded. How perfect is that to perpetuate a scam?*

Kojima obviously changed the collection dates of butterflies so that they'd appear to have been gathered before CITES went into effect. It made any scientific data on his part questionable.

"Sweet dreams. I talk to you tomorrow. Okay?" Kojima asked softly.

"I'll talk to you tomorrow," Newcomer promised.

This time, Newcomer wisely turned off his computer for the night.

First Shipment

"How does one become a butterfly?" she asked. "You must want to fly so much that you are willing to give up being a caterpillar."

—Anonymous

Butterflies are under assault, no question about it. They're at the mercy of many predators, though none are as dangerous as man. Butterflies have evolved survival techniques over the centuries but are helpless when it comes to the problem of habitat destruction. Because of this, butterflies, along with hordes of other species, are being wiped out.

Butterflies are sensitive indicators as to the overall health of the environment. They're the aboveground equivalent of canaries in the coal mine. What they're telling us is that we have to worry about our future as well as theirs. More than half of the nation's endangered butterflies are California natives, and their ecology is falling apart.

The Xerces blue, North America's first butterfly to become extinct, in 1953, was wiped out when San Francisco expanded and destroyed its habitat. It was a wake-up call. Since then a good portion of California's 250 butterfly species are on the verge of disappearing. One of the hardest hit is the tiny Quino checkerspot (*Euphydryas editha quino*). Just over an inch in size, its wings are a patchwork of brown, red, and yellow spots. The butterfly, as described by the Center for Biological Diversity, is "four engines out and about ten seconds to impact." The

Quino has slid from being one of Southern California's most abundant butterflies to tottering on the edge of extinction. The reason for the rapid decline is that 95 percent of its habitat has been lost to housing and commercial development. Due to this, the Quino is now down to 1 percent of its original population.

The USFWS's response has been to create what's best described as a "fail-safe" program. Two to four hundred Quino checkerspots are reared every year in a breeding facility run by Dr. Gordon Pratt. The only problem is that the program has become a political hot potato. A developer sure as shooting doesn't want Quino checkerspots released and flying onto land that he's bought on speculation. An endangered species found living on his property would bring development to an abrupt halt and be the source of financial heartache.

As a result, the breeding of Quino checkerspots has become an exercise in futility, since none are released to help the struggling wild population. Each year they're bred, raised, and die in captivity. Adding to the insanity is that dead adult Quinos can't be sold to collectors to help raise funds to run the program. And because they're an endangered species, dead Quino checkerspots can't be thrown away. Instead the butterflies are stored en masse in empty yogurt containers piled three rows deep on walls lined with shelves. The room is reminiscent of the last scene in the movie *Raiders of the Lost Ark* as Dr. Pratt fills yet another container with dead Quinos.

"It's extremely frustrating. I've thrown my hands up a number of times saying I'm going to give up this craziness," he admits. "We should be releasing this stuff and helping the native populations."

So far the government has refused to give permission, as butterflies continue to fight against the odds on all fronts.

NEWCOMER WAS IN HIS FISH AND WILDLIFE OFFICE early the next morning when his cell phone rang. He answered, never knowing whether it would be Kojima or one of his roller-pigeon contacts on the line. He heard Kojima's thick accent.

"I couldn't reach you on Skype," Kojima complained, in place of hello.

He was becoming increasingly like a suspicious wife. "No, I'm on the road," Newcomer replied, and softly closed his office door.

"I just want to make sure that I have your order right. You want three pairs of *hospiton* and one pair of *chikae*?" Kojima asked.

Newcomer wouldn't have dreamt of turning them down. Both species were Appendix I. "Absolutely," he confirmed. "I guess I'll also need to buy four *Morpho cypris* to use as decoys. I'll wire the money to your account in the next day or two. Is that okay?"

"It's no problem. Can I talk to you tonight, darling?" Kojima asked sweetly.

Newcomer had to laugh in spite of himself. "Sure. I'm going to my tae kwon do class, but maybe after I get home, around nine-thirty."

"Please call me," Kojima cooed, though it was more of a command than a request.

Newcomer promised that he would and hung up as Marie Palladini opened his office door and peeked in. He smiled to himself. He already knew why she'd stopped by. He'd been filling her in on his conversations with Kojima, and she was anxious to hear the latest outrageous comments. The plot was better than a soap opera.

There'd never been a case quite like this before. Palladini was convinced she'd picked the right agent, especially now that things had taken such a bizarre twist. Most other agents were too macho to have dealt with the situation. There were no such worries with Newcomer. Palladini knew that he wouldn't drop the ball but would continue to develop the case and run with it.

"All right, here's what happened last night," he began, as if it was storytelling hour.

Palladini sat entranced as she got her latest daily fix. Then Newcomer went to check on deliveries.

Kojima said the first package is due to arrive today. The phrase repeated in his mind with every step that he took. *Please let it be there. Please let it be there,"* Newcomer prayed over and over.

Even so, he scarcely believed it when he spotted the Express Mail envelope from Japan. The knife trembled in his hand as he slit the package open.

A small plastic container was snugly packed inside. He lifted it, cut the wrapping tape, and carefully opened the lid. Plain cabbage butterflies lay like tiny wisps of clouds, each in its own glassine envelope. He quickly removed them, his anticipation rapidly building. His fingers felt clumsy as a child's digging through a box of Cracker Jacks in search of the hidden prize.

Where are the other butterflies? This can't be all that Yoshi sent! Was Kojima possibly screwing with him?

He counted out a dozen cabbage butterflies before he finally found what he'd been looking for. Buried on the bottom were two clearly marked *Ornithoptera victoriae,* their wings carefully folded as though they were simply asleep. He couldn't help but think how beautiful they were.

They'd be even more so if they were alive, he reminded himself.

The real kicker was that Kojima hadn't even bothered to falsify their identity. What the heck was that all about? Newcomer put it out of his mind as he tagged the box and marked its contents as evidence. It had finally happened. The case was now fully under way.

He checked his e-mail to find a note from Kojima later that day. It was an invoice for one pair of *chikae,* three pair of *hospiton,* and four male *Morpho cypris.* The bill came to $3,080 and was signed by Kojima and Ken.

Man, this guy isn't wasting a moment, Newcomer thought to himself. *And what's with Ken's name on the bill?*

It was becoming evident that Kojima's son truly was an integral part of his business.

Sam Jojola stepped into his office as Newcomer printed out the invoice. "Don't forget that we have a garbage run tonight," Jojola reminded him.

That was right. He was so busy that he'd nearly forgotten about it. Newcomer dashed off a quick e-mail to Kojima.

> *Sorry Yoshi, but I won't be able to skype with you tonight. I'm going with a group from my Taekwon-do class to see a movie. By the way, I received your package today. Very nice, but the Ornithoptera Victoria were clearly marked. It's a good thing the package wasn't checked by Fish and Wildlife or I would have had to refuse it.*

It was an odd mistake for a seasoned smuggler to make, but he kept that thought to himself.

> *I also got your invoice and will transfer the money after I pick up a payment from a customer. I'll skype with you tomorrow night and am really looking forward to when you come to LA.*

He hit the Send button. Then he headed off for his evening garbage run. He still had Operation High Roller to juggle.

Newcomer's life was becoming more hectic as his nights and days merged into one. There were more surveillance tapes of hawks being killed, and more e-mails and Skype calls from Kojima.

He arrived home late that night to find an e-mail message from Kojima waiting for him. The *Ornithoptera victoriae* that he'd received cost one hundred dollars a pair. How many more of them did Ted want? Kojima would send a maximum of ten pair. Just transfer another thousand dollars.

I'm so happy. One of my indra martini emerged this morning, Kojima informed him.

The protected Grand Canyon butterflies would make him lots of money. Would Ted please send him a detailed map of Coconino County, Arizona? Kojima would gladly mark areas where Ted could collect *indra* butterflies for him.

Newcomer tried to contact Kojima the next evening, without any luck.

> *I looked for you on Skype tonight but you must be busy at the spa.*

Would Kojima send a picture of the *indra martini* that emerged? He'd also mailed the map that Kojima wanted and couldn't wait to see all the good collecting spots that he marked.

Who knew? It might help Newcomer track down some of the dirtbags who were collecting illegal butterflies for Kojima.

> *I also have a surprise for you next time you see me on computer camera!*

That ought to get him to Skype me back, Newcomer thought.

He didn't have to wait long. Kojima called his cell phone the following day and demanded to know where he was.

"I'll be home in about thirty minutes," Newcomer told him.

"Oh, good! In the meantime, I'll send you another butterfly photo," Kojima replied happily.

I must be his best customer, Newcomer reflected.

Kojima appeared on-screen in the same green checked shirt and navy shorts that he'd worn during the last few calls. *You'd think Yoshi would spruce himself up a bit with the way he's been hitting on me,* Newcomer mused.

He didn't have much time to dwell on it as Kojima noticed something different about Ted Nelson.

"You shaved your mustache!" he exclaimed in delight. "This is so handsome. It's much smarter. You're very nice-looking now."

"Thanks. That's good," Newcomer replied, somewhat embarrassed by his gushing.

"I wanted to let you know that I have a *meridionalis* available. It's a very special butterfly," Kojima told him as if sharing a secret. "This subspecies comes from Papua New Guinea and is more and more difficult to get. Not so many people have. My net is five hundred a pair."

"And what will you charge me?" Newcomer asked.

"Only fifty bucks more."

In truth, the *meridionalis* was Newcomer's favorite butterfly. A beautiful iridescent green, the Appendix II species is tinged with gold and found only in a handful of locations. It's tetragonal hind wings taper into a single pair of spiky threadlike tails whose delicate filaments dangle as fine as slender earrings. There was something so dainty and fragile about this smallest of birdwing butterflies that Newcomer found himself bewitched by it.

"Of course, I can't make for you a CITES permit," Kojima added.

"Why is that?" Newcomer asked.

"Because Japan has no CITES system," Kojima lied.

"They don't?" Newcomer inquired, knowing perfectly well that it wasn't the truth.

Kojima offered a convoluted explanation. "No, we can import for the CITES butterfly, but we can't make for CITES because the government isn't making the legal paper."

It wasn't worth the bother to challenge Kojima. Instead, Newcomer said that he was leaving in the morning to visit his mother.

"Your mom is not well?" Kojima asked with a note of concern.

"No, she's just lonely," Newcomer replied.

That seemed to strike a chord with Kojima. "You know, one day my brother came and took my mom to his place. It's easier that way, but I think my brother wants this house. That's why he takes care of my mommy. He needs money."

"So he wants to take your house, huh?" Newcomer asked quietly.

Kojima feigned indifference. "I don't mind. I don't want to care for this place."

"But you make a lot of money with your butterflies and antiques. You should be able to afford it." Was Kojima just cheap, or was he lying about the money that he made?

The mention of his wealth seemed to cheer Kojima right up. "Yes, I make a lot. My client just gave me thirty thousand dollars the other day for some of my antiques. Sometimes people buy an expensive butterfly that is almost ten thousand dollars," he bragged. "American customers are the best. So rich! My partner is a German guy. He's a big queen in Europe and sells lots of butterflies in the United States."

He was also Kojima's contact for the CITES I *homerus* butterfly, Kojima explained, and was going to Jamaica at the end of summer. He would probably pick some up at that time, though they'd be very expensive. Newcomer could buy one then if he liked.

"He gets them without getting caught? That's pretty scary. How does he do it?" Newcomer asked.

"I don't know how he gets them. He won't tell me," Kojima admitted.

"Those are endangered, right?" *Come on. Confess!*

"Uh-huh. They're CITES I," Kojima helpfully complied.

"Those are the most protected butterflies you can get," Newcomer concurred. "Does your partner have a Web site?"

Kojima gave him the URL, and Newcomer promptly logged on.

"Hmm. He has some interesting butterflies on here, but I don't see anything about a *homerus*," Newcomer commented, half to himself.

"That's because it's one hundred percent illegal. He make a mistake, he go to jail also," Kojima explained.

"So then how do customers in the U.S. know that he's got them?" Newcomer probed. This was one of the most secretive businesses he'd ever known.

"He's dealt with those clients for a long time. We be very careful selling," Kojima stated.

Newcomer hoped that Kojima wouldn't be quite so prudent when it came to selling to Ted Nelson.

"I almost forgot to tell you. Some *Bhutanitis lidderdali* just arrived. Do you want this butterfly also?" Kojima asked.

"How much do you sell those for?" Newcomer inquired.

"Twenty-nine dollars for the big ones. Big one is always better, right? Could you show me your big one?" Kojima teased.

Both men laughed.

"My specimen of this one is very gorgeous. No one compares to my piece."

"Really?" Newcomer responded.

"Really. They are beautiful. They look like . . ." Kojima paused and gazed longingly at Newcomer through the webcam. "Looks like you."

Their relationship had just taken another twist. *I think he's actually falling for me,* Newcomer suddenly realized. That knowledge made his deception feel even more treasonous. At the same time, it also gave him an advantage.

"Hey, I'm sorry if you had a hard time reaching me today. I was on the phone with my mom a lot," Newcomer said, hoping to deflect what Kojima had just revealed.

"How long will you be gone?" Kojima asked in a subdued voice.

"I'll be there until next Wednesday. So, I should be back on June 22."

"What can I do for four or five days by myself?" Kojima asked plaintively.

Newcomer hadn't known how much Kojima depended on him for company. It made him all the more uncomfortable.

"Well, I guess you'll just have to go to the gym a few extra times," Newcomer said, trying to pass it off as a joke.

But there was no denying that Kojima was a lonely man.

Newcomer's landline rang, and Kojima abruptly went on the attack.

"Is that your girlfriend again?" he asked testily as Newcomer answered the phone.

"It's my mom, Yoshi. I'm going to have to get off. Listen, maybe I can use my cousin's computer and send you an e-mail, but I won't be able to Skype."

Kojima waved his hand dismissively. "Don't worry. I'm just joking. I must do by myself."

He reluctantly logged off, his eyes never leaving Newcomer until the screen went black.

Widening the Net

Not quite birds as they are not quite flowers, mysterious and fascinating as are all indeterminate creatures.
—Elizabeth Goudge

Newcomer didn't visit his mother immediately in the morning. Instead, he first stopped by the U.S. Fish and Wildlife office. Kojima's second shipment was due to arrive that day, and he had every intention of being there for it. The Express Mail package appeared on time, just like clockwork, and Newcomer began to feel better after last night's heart-to-heart.

Just keep your eye on the prize, he reminded himself.

Kojima's packages were coming into the United States through the San Francisco Mail Facility. Newcomer had alerted both Fish and Wildlife inspectors and U.S. Customs to be on the lookout for them. So far, both deliveries had made it through without any problem, but then it wasn't very difficult. Mail flooded into the country like a tsunami, making their discovery nearly impossible. It was far better than any game of chance. The odds were definitely in Kojima's favor.

Newcomer studied the envelope. Kojima had listed the declared value as thirty-nine dollars.

That's a good one, Newcomer chuckled. He knew for a fact the contents were worth more than three thousand.

He slit the envelope open. Once again, the butterflies were packed in a rectangular Tupperware container with Ted Nelson's name clearly written in red ink on the lid. He proceeded to cut the wrapping tape on the container, open the lid, and remove the thick top layer of cotton.

If Nelson was really a novice, this would have been the smuggler to learn the tricks of the trade from.

Twelve look-alike butterflies were individually wrapped in wax paper envelopes. Six were the endangered Appendix I *hospitons*. The remaining butterflies were unprotected Japanese *machaons.* Were it were not for Kojima's secret code, the two species would have looked nearly identical. No inspector would have been able to tell them apart with the way that their wings were folded. Two small envelopes lay buried beneath the pile. Those contained the pair of Appendix I endangered *P. chikae.*

A relic of the Ice Age, the butterfly survives in the Cordillera Central mountains of Luzon, Philippines, and was discovered only about forty years ago. The *chikae* is fittingly referred to as "the Luzon Peacock." It looks as though tears of luminescent green splattered onto its black wings, while the hind edges are outlined in a rainbow of colors.

Newcomer's pulse was slow and steady as his heart beat with the knowledge that the case was moving forward. All he had to do was not let it spin out of control. He now had two species of CITES I butterflies to add to the growing charges against Kojima; but the law is a funny thing. It didn't matter that Kojima's victims were endangered, for one simple reason: They weren't human. Rather, they were bugs with wings. Kojima's penalty at this point would only result in a fine and a verbal slap on the wrist. The public needed to understand the seriousness of the crime. For that he'd need to land the ultimate prize. It would take the Queen Alexandra to knock the king of butterfly smugglers off of his throne.

Newcomer returned home on June 22, sat down at his computer, and promptly went to work.

Hey, Yoshi. I'm back. Will call you on Skype in 1.5 hours.

He hit the Send button, and the e-mail raced off into cyberspace.

Kojima was ready and waiting when Newcomer logged on. He wore his usual green checked shirt as he peered closely into the webcam. He resembled an owl in his wire-rimmed glasses. Kojima immediately picked up where they'd left off, as if no time at all had passed. He was cocooning five hundred *P. indra kaibabensis* and had already received six hundred e-mail offers.

"So many people want! Once they emerge I'll sell them to Japanese collectors for a thousand dollars each. I make so much money."

Things were back to normal.

"I can't wait to start collecting for you," Newcomer responded.

"Yes, but you must be very careful when collecting in national parks. Rangers are sneaky guys. They'll hide behind bushes and catch you. Then you'll go to jail," Kojima warned.

What do you know? Kojima actually seemed to care about what happened to him.

"I certainly don't want to go to jail," Newcomer replied.

"That's because you're a scared guy," Kojima taunted.

Oh boy. It never failed. Kojima always ended up turning on him.

"Well, *you* don't want to go to jail, do you?"

"I could," Kojima retorted nonchalantly.

"Oh yeah? Well, after smuggling these *hospitons* and *chikae,* you're probably a wanted man now," Newcomer cheerfully stated for the camera.

Kojima smiled as though it was a compliment.

Today Newcomer's mission was to go fishing, and he now cast his net a little wider. "Hey, does your father-in-law help you much? Or does he send stuff to people for you?"

Kojima looked at him as though he'd lost his mind. "Who? My father-in-law?" he asked, sounding confused.

"You know, your wife's father. What's his name?" Newcomer clarified.

"Oh, Charles Hanson? Yes, everything he can help me. He's a rich older guy with nothing to do." Kojima's face broke into a wide, beaming smile. In fact, Newcomer had rarely seen him so happy. "He has six thousand million dollars, and just wrote a will. He's going to divide it all between my ex-wife, my boy, and me. That's why I have to go back to the U.S. and talk to him."

Six thousand million dollars, huh? And he's leaving Yoshi two thousand million? Yoshi's living in la-la land. "That's great," Newcomer replied. "So he sends butterflies to customers for you?"

Sometimes customers were afraid they'd be caught if Kojima sent the package. In that case, he'd send the butterflies to Hanson, and his father-in-law would forward them to the client. Hanson then also handled the money.

"But my boy helps me, too," Kojima added.

"Yeah, I noticed Ken's name was on the last shipment you sent me," Newcomer replied. "Why was that?"

Kojima smiled slyly. "Because that package contained endangered butterflies. My son isn't a legal resident of Japan, and so he won't get in trouble if Japanese Customs discovers it. If they come here, I'll just say that's my brother and they'll destroy it. Besides, Ken is working as a guide in Central America for National Geographic right now, so U.S. and Japanese Customs can't find him."

Yoshi has every angle covered, Newcomer realized. "So if Customs finds endangered butterflies in one of my shipments, I just refuse it?"

"That's right. You just say, I don't order that kind of butterfly," Kojima instructed.

"Yeah, but that's a lot of money for customers to lose," Newcomer said.

Kojima gave a diffident shrug. "That's okay for me."

Sure. Yoshi doesn't have a thing to lose. No wonder he gets all of his money up front. "You know, I was thinking maybe I should contact your German partner directly about getting that endangered *homerus.* Would that be all right?" Newcomer asked. Time was ticking away, and he needed to get as many charges racked up as possible before Kojima made a trip to the States.

The lingering smile on Kojima's lips instantly vanished, and Newcomer realized the problem. Kojima wanted his cut. Kojima confirmed his suspicion.

"My partner will send a *homerus* to you for twelve hundred dollars, where I'll sell to you for seven hundred."

"Well, can I buy it through you then?" Newcomer asked.

Kojima pursed his lips in a pout. "It depends."

The money must have won out over the insult, and he relented. "I'll ask him for one for you."

Terrific! If he nabbed a *homerus* and a Queen Alexandra then Kojima would definitely spend time behind bars.

"When will you know?" Newcomer pressed.

Kojima's pupils turned into two tiny darts behind his lenses. "How come you're in such a rush?"

"What do you mean?"

"Why you want it immediately?" Kojima asked suspiciously.

"I don't want it immediately. I just want to let my client know," Newcomer said, deftly backpedaling.

Kojima shook his head in annoyance, unhappy with the response. "Everybody scared because *homerus* is Appendix I."

This guy is in need of serious medication. His mood swings are all over the place. "I'm not in any rush. I'm just curious is all," Newcomer tried to convince him.

"If you wait, I can supply. If your customer is in a rush, then something is wrong," Kojima pronounced. "Besides, my partner will wonder why you contact him when you're my client. Maybe this customer is really Fish and Wild and is trying to set all of us up."

"No way," Newcomer replied, but Kojima jumped right back in.

"I know Mr. Wallace, a Chicago man, caught twice with CITES butterflies. Now he has a big problem."

Catching a buyer was unusual. It was difficult to prove a butterfly had been smuggled in once it entered the country.

Newcomer pounced like a cat upon catnip. "Wallace? Wallace is his last name?"

Kojima promptly exploded. "How come you ask for that kind of information? You check for the names all the time!"

Newcomer was momentarily taken aback. Dealing with Kojima was harder than walking on eggshells.

"It's just that I've never heard of him before," Newcomer said, cautiously retreating.

Kojima busied himself with some butterflies on his desk, but there was no question that he was miffed. "Anyway, I'm getting the *meridionalis* if you like. Only the price is more than I thought. I pay seven hundred ninety a pair."

How interesting. Yoshi said just the other day that he'd paid five hundred for them. Is this my punishment for making him mad? "So how much more would I have to pay for them now?" Newcomer asked.

"One hundred dollars," Kojima replied smugly.

"Then the pair will cost me eight hundred ninety. They aren't legal, right?"

Kojima clucked his tongue. "That's not a problem. Look, you see?" He held a yellowed CITES permit up to the webcam. "It's just a little bit old."

Newcomer nearly guffawed at the understatement. Not only had the permit expired, but it had been issued for a totally different butterfly species. Kojima was one ballsy guy.

"We're illegal together," Newcomer said with a conspiratorial grin. "Let me think about it."

"I think the *meridionalis* is going to go very quick. Twenty customers already contact me," Kojima warned. "If you don't have a customer now, maybe you have your own Web site that you can put it on to sell."

"I have two Web sites. One doesn't have any CITES material on it, and the other one does. I give my customers the password after I've known them for a while." That was a pretty good story, if Newcomer said so himself.

"I must have taught you well," Kojima replied with a smile. "I can see your CITES Web site?"

Oh shit! What do I do now? "You need the password, don't you?" Newcomer asked, stalling for time.

"No, it's okay. I don't need it now," Kojima said, changing his mind.

"I'll show it to you when you come to L.A." Newcomer breathed a sigh of relief. That had been way too close for comfort.

Newcomer agreed to let him know about the *meridionalis* butterfly by the next day and promptly logged off.

NEWCOMER WAS QUICKLY LEARNING that with Kojima, nothing was ever what it seemed. That point was made when both the senior vice president of Human Resources and the vice president and associate general counsel for the National Geographic Society returned his call. There was absolutely no record of Hisayoshi Kojima having ever been employed for guiding or contractor services by anyone at National Geographic.

Unbelievable. Kojima had concocted the entire story! He was fooling everyone from friends to associates to law-enforcement authorities around the world.

As for the *meridionalis,* Kojima again changed his mind about their price the very next day.

"Sixteen U.S. dealers have contacted me because they can't get it anywhere else."

Dealing with Kojima was becoming nearly impossible. "Well, are you going to sell it to them or to me?" Newcomer asked impatiently.

"How much can you pay?" Kojima countered.

"It's really big, right?" Newcomer checked.

"Yeah, looks like yours," Kojima teased.

"Yeah, right. You wish you knew," Newcomer volleyed. "Will you sell it to me for a thousand dollars?"

"One thousand dollars is okay," Kojima agreed.

You bet it is, Newcomer thought. That was $450 more than what he'd originally agreed to sell it for. Kojima was nothing less than a bandit.

"You know, I really like some of the butterfly photos on your Web site," Newcomer commented. "I'm going to save some on my computer."

"I send you a naked photo if you like," Kojima offered coyly.

"You can send me whatever you want," Newcomer said, playing along.

Kojima promptly opened a file on his computer. "Maybe you like this. Oh my goodness! I open this one by mistake," he said with a mischievous giggle.

It was enough to pique Newcomer's curiosity. "What did you open?"

"Don't ask. Oh my goodness. Not this picture, either! I want to show you something nice," Kojima said winsomely, and began to move his webcam around the room.

Kojima's bedroom was larger than Newcomer had imagined. Butterfly-rearing equipment cluttered every inch of the floor, the bed was an unholy mess, and shoji screens were strewn all over.

What the hell does he want to show me, anyway?

The next second Newcomer panicked as his imagination began to run wild. What if Kojima's nude son was in the room? Or maybe Kojima was into kiddy porn and that was his next surprise. Wouldn't that be a kick in the rear? Then Newcomer would really be doomed. There'd be no more butterflies. Instead, he'd have to report the crime to the FBI and his case would be over. He'd be stuck working undercover for them on a child pornography investigation.

Please don't let me see anything that will blow three years of work on this thing to smithereens.

Newcomer's pulse skyrocketed as he waited to see where Kojima's webcam would land. He caught sight of a second computer on the opposite side of the room. The webcam settled on a porno film in progress featuring two adults, though the image was too grainy to tell if they were straight or gay. Still, he was grateful that the contents were at least legal.

"Now you show me a nice one," Kojima suggested.

His skin crawling, Newcomer countered that there was nothing nearly as interesting on his computer.

It was becoming more difficult for him to deal with Kojima every day. Another moment of this and he'd explode. That's where Newcomer's make-believe girlfriend came in handy. He explained that they had plans and that he was running late.

"Oh my God! You must think about something else. You always think about the girl. Why do you like so much the glory hole?" Kojima scolded.

Newcomer was at a loss for words as Kojima continued his rant. Ted Nelson spent way too much money on girls. He'd be far better off if he just spent his time collecting butterflies for Kojima.

Newcomer kept his mouth shut as he quickly logged off and then took a deep breath.

SKYPING WITH KOJIMA WAS BEGINNING to consume most of Newcomer's days and nights. He needed time to catch up with his other cases. There were mountains of paperwork on his desk, and a number of reports to be filed. All the while, e-mail and phone messages from Kojima poured in with the fury of a blizzard.

Ted, this is Yoshi. Please urgent to call when you wake up.

I need to speak to you, Ted. You must check your e-mail!

It was beginning to seem as if he never had a free minute. The more attention he paid Kojima, the clingier and more demanding the man became. Newcomer's patience was beginning to wear thin. He'd not returned Kojima's calls for one day, and now he was paying for it.

The *O. meridionalis* were on their way, and Kojima had still more butterflies to sell. His source seemed endless. Newcomer's undercover cell phone rang, and he jumped with a start. Naturally, it was Kojima. "Where are you right now?" he demanded.

"I'm just out doing some shopping," Newcomer replied while closing his office door.

"I sent you five photos of *Ornithoptera paradisea* yesterday and asked how many you want to buy. Did you check yet?

Newcomer rubbed his brow, his head beginning to swim. It felt as though he was drowning in butterflies. "You know, I think I saw those but thought they were the *meridionalis* so I breezed right past them."

"You understand me, right?" Kojima asked.

Newcomer stared at a stack of papers that didn't end and began to feel giddy. "Maybe not. Sometimes it's easier when I'm looking at you than when I talk to you on the phone," he admitted with a short laugh.

A moment of deafening silence followed.

"Okay. Forget it. Thank you," Kojima replied tersely, and hung up.

Damn! What was I thinking? Newcomer scrambled, having realized his mistake. "Wait, wait! Are you still there?"

The dial tone told him that it was already too late.

His punishment was that he didn't hear from Kojima for the next two days. When he did, it was as a general e-mail sent by Kojima to all his customers.

> *Just arrive new shipment from Papua New Guinea. Also Jamaica. All rare Papilio only we offer our special customers. Very large CITES1 stock pair perfect A1 (PNG). Two pair CITES1 Jamaica.*

The minimum starting price for each pair was eight thousand dollars.

Newcomer read the e-mail again. What was Kojima talking about? Then it suddenly dawned on him. It had to be code for the Queen Alexandra and the *homerus* butterflies!

Kojima had gotten them in and wasn't offering the butterflies to him first. Newcomer had clearly pissed him off. And what was this price of eight thousand dollars? Kojima had told him that a *homerus* would cost seven hundred.

He had to take action. He dashed off an e-mail.

> *Yoshi, I'm so sorry that I haven't been in touch but my uncle is very sick. Plus I don't think my cell phone is working. It must be the battery. Do you have a homerus?? $8,000 is a lot!!*

Kojima cared about family. Perhaps he would call. Newcomer received an e-mail in return. Kojima thanked Ted for his note and then delivered his own news.

> *I have very good orders now. Lots of shipments every day to the USA. Also Korea, Taiwan and Europe so can't skype now. I'm flying to LAX and Las Vegas soon. Right now all my friends are there. So maybe leave July 5. See you in Los Angeles.*

Newcomer read the message in dazed silence. It was the ultimate slap in the face. Kojima had made no mention of selling either the *alexandrae* or *homerus* to him. What was going on? He needed at least one of those butterflies to complete his case. They would show exactly who Kojima was and the kind of endangered species he could get.

He didn't waste a moment but wrote Kojima another e-mail. Did either of those butterflies happen to be available? He just hoped his desperation didn't show. Kojima would instantly sense it. Newcomer received his answer all too soon.

> *I tried to call and give you first choice but couldn't get you on the phone. I have about 200 VIP customers and 100 in the U.S. Those butterflies quickly sold out. I couldn't hold them for long.*

I don't believe this, Newcomer muttered to himself.

Kojima rubbed it in a little more by Skyping him that night. There had been so much excitement that the two pair of *homerus* sold for $12,000 each and the *Alexandra* for $9,800.

"My clients are very rich. They already transferred the money to me," he bragged.

"That's pretty fast. Are they in the U.S.?" Newcomer asked. He might as well gather what information he could from the fiasco.

One man lived in Los Angeles and the other in San Francisco.

"You better be careful. Are you sure they're safe? It's pretty risky, you know," Newcomer advised, all the while planning to give Customs a heads-up. Maybe they'd be able to intercept the packages.

"Of course. I'm always careful. Almost a hundred customers for a long time over ten years," Kojima assured him.

Newcomer didn't even want to think about the number of butterflies that must have amounted to.

"You sent it EMS?" Newcomer double-checked.

"Yeah, EMS direct. They never open," Kojima gloated.

"So where did you do the auction?" Newcomer inquired.

Kojima explained that he'd sent an e-mail to his customers with a minimum starting price of eight thousand dollars. Whoever responded with the highest bid automatically won.

There was no way not to be impressed, though Newcomer hated what Kojima did. The market for rare butterflies was even better than he had imagined. About forty people had bid on the *O. alexandrae* and sixty on the *P. homerus*. Even so, $33,000 was a lot of money for three pair of butterflies.

"So you don't have any left, huh?" Newcomer asked. *Damn, if I'd gotten just one of these butterflies, my case would be made.*

"No. That kind of butterfly lasts only one or two days, and the packages already went out," Kojima told him.

Newcomer had to rack up whatever charges he could before Kojima came to Los Angeles. It was the only way to strengthen his case. "Oh, well. I'm buying a brand-new condo and don't have that much money, anyway. What other butterflies do you still have?" he asked

Ted Nelson could be so slow-witted that it sometimes tried Kojima's patience. "You're not understanding. I already sent a message. You must check your e-mail. That's what I have in stock," he said in exasperation.

Why can't he just tell me and be done with it? Newcomer thought irritably as he logged on to his e-mail. He found a message from Kojima with photos of three different Appendix II butterflies attached.

"Okay, I'll buy all three of these. How much am I paying for them?" Newcomer asked, determined to bolster his case.

"Twenty-three hundred," Kojima replied without pause.

Newcomer agreed before Kojima had a chance to change his mind. Naturally, none of them had valid CITES permits for importation into the United States. Could Kojima send them before he came over?

Kojima said that he would. Nelson just needed to send payment for the last butterflies he'd ordered in addition to these. Would he please transfer the money to Kojima's bank account right away?

That wasn't going to work for Newcomer. His funds were running dangerously low. Besides, he planned to arrest Kojima as soon as he stepped off the plane in Los Angeles.

"You're coming to the U.S. in three days. Why don't I just give you cash then?" Newcomer suggested.

"No, I need it before to buy the plane ticket," Kojima insisted.

Was he kidding? Kojima always bragged about how much money he made. "But you just sold thirty-three thousand dollars' worth of butterflies," Newcomer reminded him.

Kojima responded as though he hadn't heard a word. Newcomer

owed him thirty-three hundred dollars, and that's the amount that he demanded be transferred. It would have been easier for Newcomer to pound his head against the wall than try to reason with Kojima.

"That's fine. I'll do it tomorrow," Newcomer said, caving.

Kojima's mood promptly changed. "Really? How come you suddenly so sweet? That's unusual," he said in a cloying tone.

"Well, if I don't send it, then you won't be able to come to the U.S., right?" Newcomer cynically replied.

Kojima gleefully launched into detail about all the lewd things he planned to do to Newcomer when he arrived in L.A., and how he wanted to do them in dirty places.

So far Kojima had only been interested in looking. It seemed he'd just raised the ante. "Oh, yeah? And how much is that going to cost me?" Newcomer asked with a snort.

"One dollar you must pay," Kojima chortled. "Don't be mad about the money. I have a Los Angeles U.S. bank account. The thirty thousand for the butterflies was deposited there. That's why I need you to send money so I can buy my plane ticket."

That's good to know. He must have opened a new U.S. bank account since the last time I checked, Newcomer noted. "Which passport are you going to travel under?" he asked.

U.S. Customs had begun to check his Japanese passport. So he now used his American one when traveling to the States. Either his father-in-law or his actor friend would pick him up at LAX, and he would most likely stay with one of them while he was there. "But maybe I can meet you in Las Vegas when I go," Kojima suggested.

"That's where all the good-looking people are," Newcomer joked.

"I don't care about good-looking. Money is more important to me," Kojima countered. "How come you buy such an expensive house, anyway?"

"Because I like a fancy place and the ladies like it," Newcomer teased.

"Who cares about the ladies?" Kojima nearly spat. "Anyway, you can sell butterflies and make a couple thousand dollars a day. Then you can easy pay for your house."

It was way too simple to bait Kojima, and Newcomer was feeling rambunctious. "I have a lady friend in Las Vegas who's very good-looking," Newcomer said, taunting him.

"She has a penis?" Kojima sharply rejoined.

"Not the last time I checked," Newcomer replied jovially. "She works as a cocktail waitress."

"How come you like that kind of girl? Too much makeup," Kojima said. He sounded as though he was ready to explode. "I like a natural face on a man. Otherwise, you cannot see what's inside."

It's time to change the topic before this dissolves into a lover's spat. "Will your son also be coming with you?" Newcomer asked.

"No, Ken's in Central America right now. He'll fly to L.A. from there. He's bringing with him maybe ten thousand butterflies."

Newcomer gave a low whistle. "Wow! How's he going to get all that through Customs?"

"We have a pass. That's why we can do it," Kojima disclosed.

"Oh yeah? What kind of pass is that?" Newcomer asked, having already guessed the answer.

"National Geographic," Kojima replied jauntily.

Not for much longer, Newcomer thought.

He intended to pull the plug on that bit of deception as soon as possible.

The Queen

On taking it out of my net and opening the glorious wings, my heart began to beat violently, the blood rushed to my head and I felt more like fainting than I have done when in apprehension of immediate death.
—Alfred Russel Wallace

Some dreams are sweeter than others. Newcomer's were beginning to be filled with butterflies every night. Sometimes he managed to catch one for a moment. Then he would release it and watch as it flew away. Perhaps he was dreaming of them when his cell phone rang at 1:30 AM on June 30. He didn't get up but listened to the voice message later that morning.

Call me quick if you still want to buy an Alexandra.

Kojima's voice sent shivers of excitement through him. It was as if his life had somehow changed. So this was how it felt to be an obsessive collector and offered your wildest dream. Newcomer raced to Fish and Wildlife and logged on to Skype at 8:30 AM, making sure that the webcam was off.

"Hi, Ted. You just wake up?" Kojima asked.

"Yeah, a little while ago," Newcomer replied, afraid this might still be part of his dream.

"Okay, I get a message from my friend in Papua New Guinea. He has one more pair of Alexandra to sell. How much can you pay?"

This is happening so fast. It won't do me any good unless I have the Alexandra in hand when Yoshi arrives in the States. "When can I get them?"

"He must send to me first," Kojima explained.

"But how will you get it when you're coming to the U.S.?" Newcomer pressed. The thought of not being able to add them to the charges against Kojima made him feel sick.

"I can wait to come until then," Kojima offered amiably.

Perfect. Newcomer's plan was back on track. "What do you think is a fair price, Yoshi?"

"This is a very large nice one. It's gorgeous. Eight thousand is good," Kojima proposed.

"I'll tell you what. Let me call one of my customers, and I'll get back to you in about ten minutes." Newcomer's mind was already awhirl with what had to be done.

"Okay, I'll talk to you then, but this Alexandra is perfect, and very difficult to get," Kojima said before hanging up.

Newcomer knew what he meant. If he didn't take it quickly, someone else would. At this point his buy money was pitifully low. Newcomer had been allotted just ten thousand dollars in total. It was a paltry amount compared with what other federal law-enforcement agencies spent on their cases. He now had only four thousand left. *Yeah, but this is the opportunity I've been waiting for. The Alexandra will give me the home run I need to bring this case to a close.*

He knew what he had to do. He would throw himself on Kojima's mercy. Maybe Kojima wouldn't make him pay up front this time. Newcomer checked his image in the mirror, fixed his hair, and then Skyped Kojima right back.

"Hey, Yoshi. I talked to my customer, and he's too nervous to pay for it in advance. He's afraid it will be confiscated, since it's an endangered species," Newcomer said, making the story up as he went along.

"So the only way I can buy it is if I can first get it, sell it, and then pay you after I receive it."

"I see, I see. Oh, man," Kojima muttered under his breath.

Newcomer could almost feel Kojima's disappointment rolling in waves at him over the Ethernet.

"Yeah, it stinks. My customer said he would pay me for it if I had it in hand, but he's not going to give me ten thousand dollars ahead of time. It's too risky. You understand," Newcomer said in his most heartfelt voice. *Come on. Give me a break,* he prayed.

"Sending to you is not a problem, but I must have advance first," Kojima insisted.

Newcomer played his hand as delicately as one would cook a soufflé. "That's exactly what I told my customer, Yoshi. I said you're a good guy and there's no problem, that we've been getting them all along. He's just too nervous to take the chance. I really want to do it but I don't have that much money on hand. Everything went into my new condo."

"Oh, I see," Kojima muttered again, then fell silent.

Yoshi isn't happy. I wonder what his cut on this was going to be. That thought gave him hope. Kojima really wanted to make this sale.

"Maybe next time," Newcomer said, then paused. "Unless your friend is willing to send it and be paid later."

Kojima clucked his tongue in frustration. That would be very difficult. He hadn't been able to get this butterfly for a long time. Now the guy was willing to give him three pair.

Newcomer realized that might be the catch. The more *O. alexandrae* Kojima bought, the more they'd be made available to him. It was time to try to reel him in. "I'd love to be able to get it, and you know me. I'd pay as soon as I got the money from my client," Newcomer added wistfully.

The butterfly was in Kojima's court. Newcomer's pulse pounded like a pair of giant wings beating in his ears as he waited for Kojima's response.

"Okay. We can pay half and half. But it's a risk for me, too. Understand?" Kojima said. "You have no other customer to ask?"

The pressure was on. He had to handle this perfectly, or the entire thing would explode.

"Tell you what. Let me check with two other customers first and get back to you," Newcomer said, hoping to finagle the best possible deal.

"The only problem is if I'm making a lot of money, I don't mind. I can buy. You understand? But I'm making nothing now," Kojima began to complain.

Welcome to my world, Newcomer thought. Did that mean Kojima had been lying about the thirty thousand dollars? Or was he just one of those guys who liked to cry poor?

Kojima tried one last sales pitch. The butterfly had a 19.6-centimeter wingspan.

"Yeah, it's a true birdwing. That's like the size of a bird," Newcomer agreed. *Christ, the thing must be enormous.*

"Maybe if it's difficult, you can sell it later," Kojima implored. "We can risk for you and me the half and half right now. Of course, you can give me a little more because I give it to you for eight thousand dollars. That is my net," he shrewdly added.

"Let me think about it and I'll call you back," Newcomer said, beginning to feel more confident. Kojima was drifting into his net.

"It's late here and I have to go to sleep. Skype me in eight hours," Kojima instructed.

That was no problem, but Nelson had one last question. He was thinking of changing banks. "Mine charges too many fees and is terrible. What bank do you use in L.A.?"

"I'm at Bank of America," Kojima replied without hesitation.

"So that's how you get payment from most of your customers?" Newcomer was thrilled that his ruse had worked.

"Yes, either through PayPal or they deposit in my U.S. account,"

Kojima confirmed. "But PayPal is terrible. I can only take out six hundred dollars a month."

"Yeah, that's bad," Newcomer said. "So how do you get your money out of your U.S. Bank of America account?"

Kojima transferred $9,999 to Japan whenever he came to Los Angeles. Anything over that amount had to be reported to the government. "I've also decided you can keep the last two payments you owe me for now. I'll pick up the thirty-three hundred when I come to L.A.," Kojima said.

Things were getting better and better. "That's great. It will save us bank fees for the transfer," Newcomer agreed. "I'll pay you cash when I see you. Now go to sleep and dream."

"Dream what?" Kojima asked, his voice softly drifting toward Newcomer.

"I know what you dream," Newcomer chuckled.

"Why isn't your camera on?" Kojima suddenly asked. "It's not showing anything."

"I've got a video error, so something's screwed up with my camera."

"You can see me?" Kojima queried, cross-examining him.

"Yeah, I can see you, but I can't get my camera working so—"

"I want to see you!" Kojima practically screamed.

Oh, boy. He couldn't risk losing Kojima over this. "I'll get it fixed when I call you back in eight hours or so. Okay?"

"Okay," Kojima agreed, temporarily placated.

Then Newcomer started his day. He immediately sent a message to the Fish and Wildlife agents and inspectors in San Francisco.

> *Be on the alert for any packages coming from Kyoto, or Osaka, Japan.*

Next, he went to check on deliveries. Another Express Mail envelope was due to arrive that day. His anticipation ran nearly as high as a child's on Christmas Eve. This was to be his third shipment from Kojima, and the package would contain the O. *meridionalis* he had ordered. His heart beat all that much faster as he saw the envelope and opened it to find the exquisite butterflies inside.

By then it was time to jump into his Tahoe and drive to Riverside, home to Latino gangs, meth labs, entomologists, and butterflies. About fifty miles from the coast and fifty miles from the desert, Riverside is known as a hot spot for lepidopterists. Newcomer headed to the University of California at Riverside, where he had an appointment with entomologist Dave Hawks. He showed him all the butterflies that Kojima had sent so far, and Hawks positively identified them as being the real deal. In fact, there were some butterflies that Hawks had never seen before. The reason was simple: They weren't available to the scientific world. That was all Newcomer needed to know.

He rushed back to the Fish and Wildlife office and was on Skype with Kojima again by five-twenty that evening.

"I can't see your beautiful body," Kojima immediately complained.

"I know, I know. You're just going to have to use your imagination," Newcomer joked.

"Too much imagination last night," Kojima grumbled.

Newcomer just hoped that Kojima would spare him the lewd details.

"When can you get your half of the money to me? I have to transfer it to my friend. Otherwise, he cannot hold the butterflies," Kojima explained.

"The earliest I can get it to you is on July fifth," Newcomer said, hoping to stall for as long as possible.

That wasn't fast enough. "All right, I'll use my own money if you guarantee that you will pay me back," Kojima said with a sigh.

Talk about a stroke of luck! "Absolutely," Newcomer confirmed. "I'll wire the money the morning of July fifth, but won't that delay your trip a bit?"

"I must wait for the shipment, so I'll come to the U.S. at the end of July. And you guarantee that you'll pay me?" Kojima checked again. "Because this one I make no money on. Minimum for the Alexandra is eight thousand dollars."

Maybe Kojima was actually telling the truth this time.

"Why are you doing this?" Newcomer asked, genuinely curious.

"Because you didn't get the last one I sold and I came out a little bit ahead on price. Besides, you must give me beautiful things in return," Kojima explained blithely.

Of course, Kojima had it all figured out. Ted Nelson was now really in his debt. *I guess he plans to take it out in a pound of flesh.* Newcomer shivered at the thought.

"That's why I'm asking you to guarantee that you'll pay for this," Kojima pressed.

"Yeah, definitely. But I still don't understand why you don't want to make some kind of a profit," Newcomer replied.

"I give it to you because I want you to start making money, but it's only this one time," he said firmly. "Besides, I hope you will order more Alexandras from me."

That made sense. The better Ted Nelson's business, the more butterflies he'd buy from Kojima. "If I successfully sell to one customer, then I'm sure that a lot of others will want to buy them," Newcomer replied, using it as added bait.

That made Kojima so happy he chattered on about the only other topic of interest to him. "I think your camera is not working because your ding-a-ling is so big. That is why it cannot take a picture." He liked to flatter the boy. Kojima especially enjoyed making Ted blush.

"Yeah, maybe that's what it is. My camera can't handle it," Newcomer genially agreed. Kojima was insatiable.

"Anyway, I asked my friend to mail the Alexandra out quickly," Kojima added.

"Okay, and then you have to come to L.A. because I'm really looking forward to seeing you," Newcomer urged. Kojima had no idea how true that was, only it wouldn't be the welcome that he expected.

The deal was done, but there was always the chance that something could still happen. Kojima was like a loaded gun. The moment he got upset, he would explode and sell the Alexandra to someone else. Newcomer sat down and typed out an e-mail.

> *I will buy the O.A. we discussed on Skype earlier today. I can pay $4,000 in advance and another $4,000 when I receive and show it to my customer. This is the most I pay for a butterfly so far so we must be careful. I can transfer the money to you on July 5 but no earlier. Do you accept?*

That spelled out their terms. Newcomer just hoped that Kojima would abide by them.

ACTIVITY FOR OPERATION HIGH ROLLER had slowed down for the summer months. Newcomer continued with surveillance tapes and garbage runs, but pigeon flies were now on hold until the fall. It was the perfect time to try and escape for a few days. He hoped it would help to repair his relationship with Allison. All he had to do was cover his bases with Kojima. That wasn't a problem. His "sick uncle" would come in handy. He shot Kojima a quick note that he'd been gone July 1–4.

A few days away proved to be the remedy that he'd needed. Newcomer was now ready to get back to work. He immediately contacted Kojima by e-mail.

> *Hi Yosh! Did you miss me?? I just returned from visiting my sick uncle in Fresno to celebrate the 4th of July. I will wire transfer $4,000 to your account tomorrow morning. Please send an e-mail or call after 8am to confirm that we still have a deal.*

I missed talking to you on Skype for the last few days. When are you coming to LA?

That should do the trick.

Kojima Skyped Newcomer promptly at eight-thirty the next morning.

"I am trying to call you before but you don't answer," he scolded.

Things were already off to a rocky start. "Yeah, I was in the shower. I'm leaving for the bank soon so I can send your money," Newcomer replied while sitting in the Fish and Wildlife office.

There was an uneasy pause before Kojima spoke again. "This time, how can I say, is a risk for me. I don't know how much you plan to sell the Alexandra for, but can I get five hundred more?"

Newcomer's stomach tightened. Something was up. "You want five hundred in addition to the four thousand?"

"Yeah, because I have a problem. I give to you very cheap for this larger Alexandra." Kojima fell momentarily silent, and Newcomer didn't dare say a word. "You'll be making much, much money on it, right?"

Kojima had changed his mind once more and was putting the screws to him.

"So eighty-five hundred total?" Newcomer asked, his mouth turning dry.

"Yes, because I'm losing money," Kojima verified.

Newcomer had no choice. Kojima had thought it over and was pissed. He was fronting his own money and had decided he deserved more of a profit. Kojima would simply "break up" with Newcomer if he argued.

"I plan to transfer exactly four thousand dollars to you today. Can I give you the remaining forty-five hundred when I get the butterfly?" Newcomer asked gingerly, hoping for some leeway.

"Maybe you can send the five hundred first," Kojima said, remaining steadfast in his demand.

"So you want me to send forty-five hundred today then?" Newcomer knew there was no point in arguing with Kojima once his mind was made up.

"Yes, that is best."

"Okay, I'll try to do that," Newcomer reluctantly agreed. He would have to scrape together the additional money.

Kojima promptly moved on, happy to have gotten his way. He asked after Nelson's uncle. Newcomer replied that the man was seventy-five years old and not very well. Kojima sadly empathized that he was only fifty-five but felt like an eighty-year-old man.

"The body starts to go," he said, with a slow shake of his head. "You wouldn't know about such things because you're still a young boy."

Newcomer didn't know about that. He was feeling pretty haggard himself these days.

"Oh, I miss you so much!" Kojima suddenly blurted out.

The outburst took Newcomer by surprise. "What?"

"I miss you so much," Kojima repeated emotionally.

Newcomer laughed uncomfortably. "Yeah, it's been a long time since we talked."

What he didn't say was how badly he'd needed the break. The case with Kojima along with his other Fish and Wildlife work and the mounting pressure at home were beginning to take their toll.

"But I can't see your face anymore," Kojima nearly cried in despair.

"I know. I'm going to have to get my camera fixed so we don't have this problem anymore."

But Kojima wouldn't let the matter go. "How come it doesn't work? My camera working for other people today," he insisted.

"I think it's a problem on my end. I'll have it fixed by tonight," Newcomer promised. "I'll have more time to mess with it now that I'm back."

The assurance that Nelson would call that evening allayed Kojima for the moment.

Newcomer was now eating, sleeping, and breathing Kojima. He hadn't realized what he was getting into taking on two undercover cases, or that it would so affect his marriage. Work consumed him twenty-four hours a day, and Allison felt that he wasn't around for her anymore. They rarely went out on evenings or weekends in case Kojima called, and when he did, Newcomer was on Skype with him for hours. He was spending more time with Kojima than with his own wife these days, and that wasn't sitting well. Newcomer was as obsessed with Kojima as Kojima was with his butterflies. The friction was building to a boiling point.

Making matters worse was that Allison's dog, Huck, was dying. She was nose-diving into depression and had never felt more alone in her life. She resented that all of Newcomer's time was being eaten up by his job and didn't know if she could handle it much longer. If this was marriage, she might want out.

Newcomer tried to be sympathetic, but his defenses kicked into gear. He'd worked hard to become an agent, and this case could make or break his career. *This is who I am. This isn't just a job.* His slogan used to be "I work to live. I don't live to work." Now it was exactly the opposite.

The divorce rate among special agents is notoriously high. Work hours are long, and travel takes up much of their time. It's become such a problem that Fish and Wildlife offers a class for couples at the end of basic training that's taught by one of their divorced agents. The advice is to think of marriage as a joint emotional bank account. If you withdraw from it, you have to put something back. Otherwise, the relationship is bound to fail. Allison reminded Ed of that.

Newcomer's initial response was *I'm here. We're together. What more do you want?* But her words struck a chord with him.

It took more than just sitting together in a room. He also needed to be there emotionally. Allison compromised as well. "If you're not going to be around for a while, that's okay. Take some advances. Go and do your case. But when it's over, you've got to come back."

Newcomer promised that he would try.

He was back on Skype with Kojima again by 11:45 that night.

"I transferred the money to Papua New Guinea today," Kojima informed him.

Newcomer breathed a sigh of relief. He'd made it past the first hurdle. "I transferred the money to you today, also. It went into your Kyoto account in U.S. dollars."

"Oh no. The best way is Japanese yen because of the exchange rate. That just cost me a hundred dollars," Kojima complained.

"I'm sorry," Newcomer apologized. "Here I'm trying to do a good thing and mess it up."

"You always make a mistake." Kojima good-naturedly brushed it off.

"I tell you, Yoshi, I don't know why you deal with me. I'm just a walking mistake. I'll buy you a hundred-dollar steak dinner when you come to L.A.," Newcomer promised.

"There are nicer things you can do for me," Kojima boldly proposed.

"That will cost *you* more than a hundred dollars," Newcomer joked, as Kojima now began to examine something in his hand. "What have you got there?"

"It's a butterfly pupa, only a bee is hatching inside. There's something wrong with some of the pupae. That's why so many bees are in here."

Kojima wasn't kidding. Bees were flying all around the room as he began to swat at them and scratch his arms. Then he turned his attention back to Newcomer. "You miss me?"

Keep him happy, Newcomer reminded himself. "Yeah, I actually did."

"Then why don't you show me your beautiful thing?" Kojima suggested. "After all, you miss me and I miss you."

"I'll save that for when you come to L.A. in person," Newcomer said, playing his trump card.

Kojima nearly squealed in delight. "Oh my God! I'm so excited I cannot sleep! By the way, my friends are collecting more *indra* for me. This time I get *indra nevadensis.* It looks like you and is also expensive. I can't believe you want to charge me a hundred dollars for sex!" he teased.

"Hey, man, I'm worth it," Newcomer joked.

"Only if you have a huge, huge big one like my actor friend," Kojima bantered, fully enjoying the conversation. "Why don't you stand up and show me?"

"I'm wearing a T-shirt," Newcomer coyly replied.

"No, I want to see down more," Kojima eagerly instructed.

"I've got on my jeans," Newcomer quipped.

"You can just open them," Kojima urged.

"You're bad," Newcomer said with a chuckle.

"I only request because your money transfer cost me a hundred dollars today," Kojima reminded him.

"I know. I'll try to make it up to you. I'm sorry." They'd had the obligatory sex talk. It was time to get back to business. "So when are you coming to L.A.? Did you decide?"

"I'm waiting for your secret butterfly," Kojima said, referring to the *alexandrae*.

"Yeah, but when do you think you'll come?" Newcomer persisted. He had to get Kojima to L.A. It was the only way he'd ever be able to arrest him, and the entire case hinged on it.

"Once I receive the Alexandra, I ship it to you and then I come," Kojima clarified. "Anyway, I have to meet with customers in San Francisco, Nevada, and Arizona."

"Speaking of customers, did your clients ever get the Alexandra and *homerus* that you sent?" Newcomer inquired.

"Yes, already. No problem," Kojima replied.

Score another one for Kojima. That meant FWS inspectors hadn't found Kojima's packages even though they'd been alerted.

"The one package is worth twelve thousand dollars, and the other is fifteen thousand. Yet I only make five hundred dollars on my sale to you. That's why I'm losing. How much money will you make on your deal for the Alexandra? Tell me, tell me, tell me!" Kojima insisted, like a child throwing a tantrum.

"Don't worry, I'll make it up to you," Newcomer said, and then noticed something new in Kojima's hand.

Kojima was playing with a live beetle as they spoke. The bug's antennae twitched and Newcomer's stomach jerked as he remembered Kojima's apartment in L.A. The place had stunk of mothballs and insects both dead and alive. He had no doubt Kojima's home in Kyoto smelled the same way. He grew more repulsed by the minute. Bees flew all over his place, there were bugs in his bedroom, the futon was unmade, and Kojima watched porno on his computer night and day. Just the thought of all the grime and bugs made Newcomer feel itchy.

Kojima glanced up once again and began to walk away from his desk.

"What's going on?" Newcomer asked in alarm. Who knew what was moving around in his place?

"Oh, a beautiful butterfly is hatching," Kojima intoned in a mixture of wonder and delight, a look of sheer joy spreading across his face. "Can you wait? I will show it to you."

"Sure. It's got to pump all the blood into its wings first, right?" Newcomer asked, wishing he could see the equivalent of a National Geographic moment.

Kojima returned a few minutes later with a newly hatched butterfly in the palm of his hand. It was spectacular, with wings in shimmering shades of blue and green so vibrant they verged on the surreal. It was as though they'd been sprinkled with fairy dust. It took Newcomer a moment to realize that the butterfly was already dead.

"This is *maackii* from a special area in Japan and is worth about two hundred dollars," Kojima said, holding it up to the webcam so that

Newcomer could get a better look. "A little bit scratched already, but that's okay."

Newcomer's stomach twisted a bit more. "How did you kill it so quickly? Did you just pinch it?"

"Yes. I pinch your dick also the same way." Kojima broke into giddy laughter and clicked his fingers together like a dancing lobster's claw. "Show me. Show me a little bit. I can pay you a dollar," he said and waved a bill in front of the webcam. "Now will you show me your hairy bunny?"

"Not till you come to L.A.," Newcomer said staunchly.

Kojima was on one hell of a roll tonight. "Oh my God. You are so nasty. No wonder I can't sleep well. Anyway, I'm so happy because I have a hundred pupae and two more *maackii* are hatching tonight."

Newcomer's nausea settled deep in the pit of his stomach. How could Kojima profess such love for a butterfly and kill it the very next minute? *Oh, yeah. That's right. It all came down to money. The butterfly was worth two hundred dollars.*

"Do they fly loose in your bedroom when they hatch?" he asked. It was almost as if Kojima was living in a zoo.

"No, not flying. I put them in this." Kojima moved his webcam to reveal fabric containers that hung from the ceiling like Japanese lanterns. The butterflies could pump their wings inside before they were killed.

"The Alexandra that I'm sending you is bigger than the one in this picture. Looks like you and is the size of your dick," Kojima teased, and held a photo up to the webcam.

There it was again, that same maddening feeling. Part of Newcomer was psyched, while the other part was pissed. *That's one more Alexandra that won't fly free in the wild.*

"That's great. Maybe I'll sell it for more than I planned," Newcomer quipped.

"Your customer will be very happy," Kojima replied.

"Yeah, I think we're going to build a good business together, you and me. You're like the insect rock star," Newcomer said, flattering him. "You're the butterfly movie star."

"I hope so. Everyone knows I sell illegal things," Kojima gloated.

"That's true. You're the number one smuggler," Newcomer agreed. "They'll make a movie about you one day, Yoshi."

It was late and Newcomer had seen enough for the evening. "I'm going to sign off. I'm really tired and need to get some sleep."

"Do me a favor. For one quick second, can I see your beautiful thing? Then I can also sleep tonight," Kojima pleaded.

"Not until you come to L.A."

"Just for one second, please! Just show me one second," Kojima begged.

Newcomer repeated his mantra. "You've got to wait till you come to L.A. How else am I going to get you here?" With luck, the tension would continue to build until Kojima couldn't take it any longer.

"So nasty, so nasty," Kojima moaned, rocking his head in his hands.

"I already sent you forty-five hundred dollars today. That's good enough for now," Newcomer said, and called it a night.

He was exhausted by the time he logged off at 12:30 AM, but Kojima was nothing if not relentless.

A voice message was waiting for Newcomer when he woke the next morning. *Call me!* Kojima demanded.

For chrissakes, doesn't this guy ever give it a rest? There was no longer any escape. Newcomer did as commanded.

Kojima had a task for him. Some of Kojima's antiques were on consignment at a store in San Diego. Would Ted please go and pick them up? There were fifteen pieces in total, valued at twenty thousand dollars. Kojima would retrieve them when he came to L.A. "I want to see you soon, my nice-looking boy," Kojima ended the message.

Newcomer's fourth shipment of butterflies arrived that day. Three perfect *O. paradisea,* each with a wingspan of nearly 7.5 inches, were buried deep inside a Tupperware container. Newcomer stared at them in admiration. Also called the "Butterfly of Paradise," the males resembled winged emeralds with flashes of topaz in flight. Once again the package had slipped through Customs undetected. So far, it was 4–0, and Fish and Wildlife was on the losing end.

Newcomer had to leave town to teach a defensive-tactics course for a few days. Ted Nelson's sick uncle provided the perfect excuse. It seemed his ailing relative had taken a turn for the worse and finally died. Newcomer sent Kojima a short e-mail.

Hi Yosh. I picked up your antiques.

He carefully listed each of the items so there would be no mistake. He didn't want to be accused of stealing anything.

I have to go to Fresno for my uncle's funeral. Talk to you on Saturday or Sunday when I return to LA.

The excuse worked like a charm. Not only did Kojima respond with a note of sympathy but also had a special surprise for him.

Do you still want homerus? I have a pair and they're perfect. If you want I can hold them until you find out if your customer is interested. See you soon my special young nice guy.

Yoshi and Ken

Everything was working out for him.

Of course he wanted it. He'd have a grand slam if this could be pulled off. Kojima would be lucky see so much as a single butterfly outside his jailhouse window for a long time.

Newcomer's few days away flew by. He was no longer worried.

The rest of the case should be a piece of cake. He returned home on July 16 to another e-mail from Kojima.

> *I received the Alexandra today and will mail it this afternoon. Advise me when you receive and I am going to LAX.*

Things couldn't get any better. Newcomer quickly replied.

> *Great news! Hopefully the Alexandra will arrive tomorrow. I'm back from Fresno and can skype with you later tonight. When can I expect you in LA?*

He was feeling so good that he actually looked forward to speaking with Kojima. He sat down and logged on to Skype at six-thirty that night. "Hey, Yoshi. It's good to talk to you again. Can you see me on camera?"

Kojima rubbed his eyes, having just woken up. "Yes, such a beautiful boy. How's your mommy feeling after the funeral?"

"She's okay." The anticipation was too much for Newcomer to wait a moment longer. "Listen, you mentioned in an e-mail that you have a *homerus*."

"Oh, are you interested?" Kojima sounded faintly surprised.

That was a weird question. Kojima knew perfectly well that he wanted it. "Yeah, I have this one client that's interested."

"This one is gorgeous. I've never seen such a beautiful one before," Kojima began his pitch.

Newcomer got right down to business. "How much do you want for the pair?"

"Last time I got twelve, but I'll give you a special price. It's nine for both. They're in beautiful condition," Kojima replied.

"That sounds fine. How much do you think I can sell them for?" Newcomer asked.

"Fourteen to sixteen is a good price," Kojima told him.

It was too good to pass up.

"Okay, I'll check with my client tomorrow. We'll settle what I owe when you come to L.A.," Newcomer said.

"Yes, otherwise I have no money and can't go anywhere," Kojima replied.

That wasn't something Kojima would have to worry about. Newcomer already knew where he'd be going.

Then it was time for their nightly game.

"Now you show me your beautiful one?" Kojima said.

"I can't today because my mom and sister are coming back any moment. They just went to get a pizza."

This wasn't the way their call was supposed to go. Kojima slowly blinked from behind his round glasses. "You're very difficult to catch lately. How come?"

It had been a long day and a big drive home, and Newcomer was ready to grab some dinner and relax. Besides, he hadn't seen Allison in a few days and needed to spend time with her or their fragile situation would only get worse.

"I was in Fresno," he replied quietly.

"You still in Fresno last night? I tried to call you then," Kojima accused, starting to sound upset.

It felt as though a noose was tightening around his neck as he took a deep breath. He'd been bone tired when he'd retrieved Kojima's messages last night. They'd sounded urgent, but then they always did. Kojima was just lonely and wanted to talk. "Yeah, I got them, but I was dealing with my family. Besides, you called pretty late. One of the messages was around midnight, and I was already asleep."

Kojima slowly calmed down. "That's okay. I hope you can make lots of money with the butterfly I sent you."

"I'm sure that I will. I can't wait to see the Alexandra." Newcomer wouldn't believe it until he actually held the legendary butterfly in his hand.

"Looks a little bigger than your dick," Kojima said, attempting to restart their nightly game.

Newcomer gritted his teeth. *God help me. I don't know how much more of this I can stand.* "Hey, you know what? My mom and sister just walked in. I've got to go."

He quickly logged off. He'd have to hold on a little while longer. *Just keep it together,* he reprimanded himself. The case wasn't over, and anything could happen.

He shot Kojima a note the next day.

> *I talked to my customer in Palisades Park. I'll buy the homerus from you for $900. Hope you can send it before you come to LA.*

It took less than an hour for Kojima to write back. There'd been a misunderstanding. Nelson was badly mistaken if he thought a pair of highly endangered *homerus* would go for that amount. The price was nine thousand dollars. Nelson should please inform his client.

Newcomer stared at the message in disbelief. *Here we go again,* he thought. By now he'd bought nearly fifteen thousand dollars' worth of butterflies and still owed Kojima half of the money. No way in hell could he ever hope to purchase the *homerus* unless Kojima gave it to him up front.

He put all thoughts of the butterfly on hold as he rushed out to check on deliveries. This was the day that the Alexandra was finally due to arrive. It was the moment of truth that would elevate the case from the ordinary to the sublime. Did Kojima really trust him enough to have sent the world's most highly prized butterfly?

Newcomer's heart beat, his throat dry as sand, as he caught sight of the Express Mail envelopes that were piled up. He rummaged through the stack until he caught sight of the package from Kyoto, Japan. It was addressed to Ted Nelson from Ken Kojima. The Customs Declaration form stated that sixteen pieces of "Dry Butterfly" valued at thirty dollars were inside.

He removed the top layer of fourteen common butterflies as though they were so much confetti. He could barely contain his excitement. Then his eyes landed on the prize that he'd waited so long for. Hidden beneath the decoys lay two large triangular envelopes. He gingerly lifted what looked like fragile pieces of origami, his fingers trembling as they slowly unwrapped a stunning pair of *Ornithoptera alexandrae*. They were like a myth come alive. The butterflies were even more magnificent than he had imagined.

The female had rich chocolate wings dotted with spots of cream; tufts of red fur encircled her thorax. Her wingspan measured a good foot, while the male's wings were seductive blue and green velvet framing a body of pure gold. All it took was one look to understand why the butterfly was deemed so highly desirable. The Alexandra had already been harmed by habitat loss. Now rumors swirled of vaults filled with hundreds of them in Japan. The butterflies were being stowed away by commercial dealers in anticipation that they'd go extinct. Once that happened, Alexandras would be worth twenty thousand dollars a pair. Kojima and his kind were pushing the species to the brink.

I got him. I got him. The words repeated themselves in Newcomer's head as he bagged and tagged the evidence. There'd been a chance Kojima might have received probation. No longer. Now he'd definitely go to jail. There was only one thing left to worry about: blowing the case before Newcomer could get his hands on him.

He checked his cell phone to find that Kojima had already left three pressing messages. Had Ted received the package yet? It was urgent that he let Kojima know.

Newcomer chuckled to himself. It was the first he'd known Kojima to sweat, but this time his own money was on the line. *Let him worry a little longer,* Newcomer thought, and headed straight to the U.S. Attorney's Office. He left armed with an indictment for Kojima's arrest.

He didn't Skype Kojima until later that evening. If Kojima was worried, it didn't show as he sat at his computer drinking a cup of tea.

"Hi, Yoshi. How you doing?" Newcomer asked.

"Okay. I sent out twenty packages of butterflies today. It's not big money but continue selling well," Kojima said amiably.

"That's good. Hey, I noticed you put Ken's name on the Express Mail package again," Newcomer said.

"It's okay. If Customs finds the package, they can't catch him in Japan because he's not here. So they just throw it away. Anyway, you receive it no problem?" Kojima checked.

"Yeah, everything's good," Newcomer confirmed.

Kojima then advised that Nelson delete all their old e-mail messages. That way Fish and Wildlife would never find them on his computer. Newcomer agreed to be more careful.

"By the way, I'd like to get that *homerus,* but nine thousand dollars is a lot of money for me."

"You must understand, CITES I is not so cheap, and *homerus* is the most expensive butterfly. *Homerus* is more difficult to get than Alexandra," Kojima said.

Newcomer explained that he'd already asked the same client that had bought the Alexandra and there was no way he'd come up with another ten or eleven thousand dollars in the next few days.

"Oh, I see," Kojima replied, clearly disappointed. "Anyway, I must sell to other people then because I don't want to hold it for too long. I'm sorry."

Let it go. You've already got enough to charge Kojima and get jail time. Don't get greedy. Just concentrate on getting him over here, Newcomer decided. "That's okay."

"Also, it's very difficult to get you today because I wonder about whether the package arrived or not. You must give the Alexandra to your customer quickly. Don't hold on to CITES I for too long, or you could have big problem. Customs might come and want to see what's in the package," Kojima warned.

Why was he suddenly so antsy? "I'll take it tomorrow morning and try to convince my customer to also buy the *homerus*. I'll let you know if he changes his mind," Newcomer replied.

"This is my last request before you sleeping. Could you please show me something?" Kojima asked, and leaned in close to the webcam.

Newcomer shook his head in amusement. "You ask me that every time."

Kojima clapped his hands. "That's why it's so exciting. It's also why I can't give you a free butterfly yet."

Newcomer played along. "Oh, really? So what kind of free butterfly would I get?"

"It's a secret, but you can make a lot of money on a free one," Kojima said, trying to tempt him.

"Not till you get to L.A., and then I have a special surprise for you," Newcomer said enticingly.

"A surprise?" Kojima clapped his hands again in delight. "A big one is okay."

"Yeah, it'll be a big surprise," Newcomer confirmed.

"Why don't you show me now?" Kojima urged. "Just for one second."

"Not till you get here," Newcomer teased.

"You're so nasty," Kojima replied, thoroughly enjoying the game. "Anyway, I'm waiting for your beautiful big one tomorrow. Send to me a photo."

"Don't hold your breath," Newcomer said, chuckling, and thought, *If Kojima only knew.* "You have to wait till you come to L.A. to see my Alexandra."

"You promised that you'd send me a photo," Kojima reminded him petulantly.

This always happened at the end of their conversations. The playful teasing became serious, and Newcomer had grown tired of it.

"I've got to have something to get you to L.A.," Newcomer said. "Anyway, I'll talk to you tomorrow."

"I hope so," Kojima replied wistfully.

Newcomer promised.

Web of Lies

It had to be someone who knew they could never display them or sell them, but was obsessed with owning them.
— Michael Shaw

Kojima e-mailed photos of more butterflies to Newcomer later that night, but Newcomer was now fixated on obtaining the *homerus.*

This would make my case. I'd have the Big Four. The thought ate away at him until he could finally stand it no longer. Newcomer wrote and agreed to buy one of the CITES II butterflies for seven hundred dollars.

> *Send right away when you can. By the way, I'd really like to try to sell the homerus. I think I can if my customers see it in front of their faces. But I'll only buy if I can pay after you arrive in LA. Otherwise, I'll have to pass on it.*

That should do it. Let's see how much Yoshi really wants to please me, Newcomer thought, and sent the e-mail.

Kojima's response came by cell phone early the next morning. "Hi, Ted. It's Yoshi. Where are you now? At your house?"

"I'm in the car," Newcomer replied, his pulse racing.

"Oh, okay. I guess that's why I don't understand your English," Kojima joked, referring to an earlier fight between them.

"Did you get the e-mail I sent about buying the *paradisea*? Listen, I'd also like to try to sell that *homerus*," Newcomer said, not wanting to wait a moment longer. "My one customer isn't interested, but I can show it to some others and say, 'Look, here it is. Do you want it?' The only problem is that I can't pay until you come to L.A. But I'll tell you what. In exchange, I'll give you ninety-five hundred for it."

He hoped it would prove enough to sweeten the deal. He could hear Kojima breathing as he held his own breath.

"Okay. I just need to check this pair and make sure they're perfect," Kojima agreed.

"Thank you, Yoshi," Newcomer said, unable to believe this stroke of good luck.

"Okay, and I miss you so much," Kojima said, sounding heartfelt.

"I know, I know. I can't wait to see you again," Newcomer automatically replied, already anticipating the day that Kojima would arrive.

He was out on a surveillance run that night when his undercover cell phone unexpectedly rang. He answered without really thinking.

"Hi, Ted. Where are you now?"

Damn, it was Kojima. He'd have to do something to quickly end the call. "I'm hanging out at a friend's house. We're watching TV and getting ready to play poker. I've got to be prepared for Vegas."

"Oh, are you going to Las Vegas?" Kojima asked in surprise.

"Yeah, I'm meeting you there. Remember? So what's up?" Newcomer inquired, doing his best not to ruffle him.

"I just call to talk but you are busy," Kojima replied, though he sounded somewhat flustered.

There was nothing Newcomer could do about that right now. "Yeah, maybe we can talk tomorrow night," he suggested, and hung up.

He couldn't believe it when his cell phone rang again only five minutes later. Kojima knew he was available, and Newcomer had no choice but to answer.

"Hi, I have one short question," Kojima said.

"Okay. What can I help you with?" Newcomer asked, doing his best to keep his patience in check.

"Did you deliver the Alexandra to your customer?"

Kojima had to be kidding. *This* was what he was calling about? "Yeah, I delivered it today around lunchtime."

"Oh, I worry you holding it," Kojima said, clearly stalling.

"Don't worry, Yoshi. It's been taken care of," Newcomer assured him.

"Okay, thank you. Bye bye," Kojima said, and reluctantly hung up.

It was as though he was keeping tabs on Newcomer every minute.

Newcomer made sure to Skype Kojima the very next day, but oddly enough, he wasn't around. *Yoshi's probably at the gym or getting a steam bath,* Newcomer mused, and put it out of his mind.

Kojima reached him by cell phone a short time later. "I'm calling because I may have a virus with my computer."

"That's because you're downloading too many dirty pictures," Newcomer joked, having begun to feel rock solid about the case.

"Yes, too exciting. That is the problem," Kojima agreed jovially.

"Hey, I'm on my way to the beach. It's so hot it's like a hundred degrees today," Newcomer said while heading in to work. He had another surveillance tonight on a roller-pigeon fancier's house. "By the way, I won't be able to talk later because I'm going to a movie with some friends."

Kojima's mood turned on a dime.

"Daytime you're at the beach and nighttime you're at the movies. You have more time for that than for me," Kojima complained bitterly.

Newcomer hated when Kojima started on one of his rants. "We'll have to go see a movie while you're here," Newcomer offered, but Kojima was in no mood to be placated.

"Thank you very much, but I don't want to stop you from looking for a beautiful guy or a beautiful girl. I don't know which one you looking for but is okay," he said peevishly.

The conversation was so crazy that Newcomer couldn't help but goad him. "There are both of those on the beach," he teased.

It had the expected effect. "Okay, talk to you soon," Kojima said, and began to hang up.

"Hey, wait a minute. You're going to send that other *paradisea* butterfly, aren't you?" Newcomer figured the thought of more money would put Kojima in a better mood. He was right.

"Yes, but you have to wait for the *homerus*. I look and there's a scratch. But don't worry. My friend will get me another pair," Kojima said.

Damn! I really want that butterfly as the slam-dunk for the case. The only problem is I need it now. "How bad is the scratch? Is it on the wing?"

"Yes, and if they're scratched, you can't get a high price for them," Kojima reminded him.

"That's too bad. I hope you can send it before you come. That way I can sell it and give you the money," Newcomer pressed.

Kojima promised to do what he could.

"Thanks. I'll take a picture for you at the beach," Newcomer jested while pulling in to the Fish and Wildlife parking lot.

"I want a picture of your ding-a-ling only," Kojima said.

"All right, we'll talk about that when you get to L.A.," Newcomer vowed.

The prospect made Kojima even happier. "Oh my God! Okay, see you soon," he agreed.

Now all Newcomer had to do was wait. He felt so good about the case that he even took time for himself and Allison over the weekend. He celebrated by turning his cell phone off. When he finally checked it was flooded with messages from Kojima. Each was more urgent than the last.

This is Yoshi. Please call.

Ted, this is Yoshi. How can I catch you? I'm leaving soon to the U.S.

If you home please contact me. Urgent. I'm leaving tomorrow afternoon. Then I'm going to Las Vegas.

His voice was more strained with each call. Kojima finally sent Newcomer an e-mail.

> *I am going to LAX but can't contact you. So I first go to Las Vegas and maybe stay 10 days and then to San Francisco. Maybe I can see you at the end of my trip. Contact me by e-mail.*

Damn! How could he have let this happen? Not now when he was so close to finally catching him! Newcomer quickly came up with an excuse and pounded out an e-mail.

> *I am stuck in San Diego without my phone. I came here on Saturday and my car was wrecked. I can't drive until they repair the bumper tomorrow. Luckily, I have a nice friend here and can stay with him. I'll be back in LA on Monday night. Are you leaving on Monday or Tuesday? Call me when you're in LA and can see me.*

He waited a few hours for Kojima to respond but received no answer.

Please, dear God, don't let me have blown this now! He held off until he could stand it no longer and then tried calling on Skype. It felt as if a two-ton weight was lifted from his shoulders when Kojima answered the line.

"Hello?" Kojima said, sounding as though he was a hundred years old.

"Hi, Yoshi. It's Ted. How are you?"

"Okay. Where are you now?" Kojima asked, his voice hoarse and frail. Something was definitely odd. He sounded tired and despondent, as though he hadn't slept in days.

Stick to your story. "I'm at my friend's house in San Diego. He has Skype on his computer but he doesn't have a camera."

"Oh, really? I'm at the airport now," Kojima feebly replied.

"You are?" Newcomer's heart began to beat faster. It was finally going to happen. Kojima was coming to L.A. Everything was falling into place. "How does that work? How can you use Skype at the airport?"

"Oh, I just use my e-mail number. I was just speaking to some people that came in here now," Kojima said.

"Wow! So you're actually at the airport and there are other people around, huh?" Newcomer asked, still trying to figure out exactly what was wrong.

"Yeah, yeah, yeah," Kojima mumbled.

"Cool. So when are you leaving? What time?" Newcomer asked anxiously.

Kojima uttered a one-word response. "Now."

"Okay, what time do you get into L.A.?" Newcomer continued, his mind beginning to race.

"In the afternoon," Kojima replied.

"Tomorrow?"

"Yeah," Kojima said, then corrected himself. "No. Your tomorrow."

Kojima would arrive on Monday. That would work for Newcomer. "Do you know what time?"

"Oh, is . . . I didn't check yet. I don't know but I think you will not be here because my son is calling," Kojima said, seeming to dodge the question.

Newcomer had no idea what Kojima was talking about. "So who's picking you up then?"

"My son. He got your message because I didn't check for my e-mail yet," Kojima explained.

Nothing he said seemed to make sense. It didn't matter. The important thing was to get Kojima to L.A.

"Yeah, my car got wrecked, and I can't drive until tomorrow. So I'm stuck in San Diego right now. Listen, we need to meet when you're here and go over how much I owe you," Newcomer said, using that as bait.

"Oh, is over ten thousand dollars," Kojima replied right away. "I didn't bring much money, so I'll need it immediately."

That's the plan, Newcomer thought grimly. "No problem. I'll give it to you as soon as I see you."

Kojima hesitated. "Yes, but I must go to Las Vegas."

"You have to land in L.A. first," Newcomer reminded him.

"Yeah, but I can't meet you," Kojima said.

The excuses were mounting. Kojima no longer seemed all that anxious to see him. "Okay, then when are you coming back to L.A.?"

"Maybe the end of August," Kojima morosely responded.

"The end of August!" Newcomer was stunned. *What the hell is going on?*

"But only one day can I meet you," Kojima added. "I left a message yesterday."

If Kojima was playing hard to get, he was doing a good job.

"So, you're going to spend, like, two weeks in Vegas?" Newcomer checked.

"Yeah, something like that," Kojima replied. "Because my friend will take me to the Canyon for butterflies. That's why I must go."

"That's great," Newcomer said, pondering his next move. "So you're going to land in L.A. and then just catch another flight to Vegas?"

"Yeah, a couple hours in transit," Kojima confirmed.

Newcomer would have to find a way to stop him. "What flight are you coming in on from Japan?"

"United, but I don't know the number," Kojima replied.

That was possible, but it could also be a major line of BS. "Are you sure you don't know what time you get in tomorrow?" Newcomer asked, desperate to pin him down.

"I didn't check yet because my son arranged for the ticket," Kojima said, purposely keeping things vague. "Oh, I want to let you know I also have more butterflies that you can buy. Two male and one female *O. goliath crocus*. Huge big ones. Very rare, for six hundred dollars."

"Yeah, that sounds great, but it's too late because you're already at the airport, right?" Newcomer replied.

"No, this one is already packed. My son will just drop it off at the post office," Kojima said, beginning to sound like his old self again.

Just buy the damn things and make him happy! "Okay, that's a good price. I'll take them. Now, when am I going to see you?"

Kojima's voice once more turned heavy. "Oh, I try to go back to L.A. for a few days, but I don't know."

"Well, after tomorrow I'll have more—" Newcomer began, only to have Kojima cut him off.

"No, no. Tomorrow you are so busy then maybe I cannot see you," Kojima demurred.

This was crazy. Could his case actually be unraveling? "Well, don't go back to Japan without seeing me!" Newcomer practically begged.

"But I don't know if you have time," Kojima replied, sounding further deflated. "You always at a movie or something. If I cannot see you, then there is nothing for me to do."

So that was the problem! He should have realized it all along. Kojima felt neglected. Newcomer quickly scrambled to make up for lost ground.

"If I know you're going to be here, then of course I'll spend time with you. Listen, do you have a list of everything I owe because I really want to pay you." Even if Kojima was angry, he would still want his money, Newcomer reasoned.

"Maybe you'll pay me, or maybe you can't see me. I don't know. But

I only have a credit card and a few hundred dollars," Kojima pitifully replied.

Newcomer had to think fast to save himself. "Yeah, I need to pay you then. Maybe I can come to Vegas and see you. Just call me after tomorrow and let me know where you're going to be," he urged.

"Okay," Kojima muttered.

Is nothing going to jolt him from his depression? "I'll just drive over there," Newcomer repeated again.

"Okay, I'll let you know. Thanks. Bye bye," Kojima replied.

But there'd been an odd finality to the call. Something had been weird about the entire conversation. *What was it?* Newcomer wondered, racking his brain. Then he realized what it was. It had been the eerie sound of silence. The background had been deadly quiet. There'd been no din of people stirring about or the slightest hint of a flight announcement.

Damn the man! Newcomer's pulse pounded as he quickly Skyped Kojima again.

Kojima sounded even more surprised to hear from him this time.

"Hey, Yoshi. I was just thinking. What about all your antiques that I picked up?" Still no sound in the background.

"Antiques? Oh, I forgot about those," Kojima replied, as though in a daze.

"Well, you're going to want to take care of that while you're here, aren't you?" Newcomer asked.

Then the seemingly impossible happened.

Kojima's webcam briefly flashed on. Kojima must have hit the camera button by accident. It just as quickly shut off. However, it had been on long enough for Newcomer's worst fears to be confirmed. The heavinesss he'd felt now turned into a four-ton weight that crashed down upon him.

"You're not really coming to the U.S. after all, are you?" Newcomer asked hesitantly.

"What do you mean? I'm leaving today," Kojima replied warily.

He's played me long enough. Yoshi is the ultimate con man, and God help me if I haven't become his mark. Newcomer couldn't do anything other than to laugh helplessly. "Yoshi, your picture just flashed, and you're not at the airport at all. I saw you at your house."

"That's impossible," Kojima said, bristling. "This is Japan Airline office."

Did Kojima think that he could actually keep lying to him? It was time for a reality check. "No, your picture clicked on for a split second, and I just saw you there in your room."

"No! How can you see?" Kojima fumed, caught between anger and embarrassment.

The guy was unbelievable. "I know what I saw," Newcomer insisted, standing his ground.

Kojima flew into a rage. "I can't discuss it! I'm not flying yet."

Warning bells and sirens went off in Newcomer's head. *What am I doing? I know better! This is a bad mistake! I've got everything to lose. I have to turn it around somehow.*

But the case he'd worked so hard on began to topple like a row of dominoes.

"Well, I hope you're coming, so . . . I really want . . ." He fumbled for the right words.

"Okay. Thank you," Kojima replied brusquely, and swiftly clicked off.

"I really want to see you," Newcomer finished, but it was already too late. Kojima was no longer there. Newcomer's words landed dully in the Ethernet and were swallowed up in space.

The Chase

I'll be floating like a butterfly and stinging like a bee.
— Muhammad Ali

What do I do now?

Panic set in as Newcomer kept trying to Skype Kojima and Kojima refused to answer. *Okay, the third time is a charm,* Newcomer told himself, and tried once again. The line rang and was suddenly picked up. "Yoshi, are you there?" Newcomer asked frantically, but there was no response.

"Hello?" Newcomer ventured.

The next sound was that of a bubble softly popping underwater as Kojima hung up.

Let's try once more, Newcomer resolved. He called and the line clicked on again. "Are you there?" he asked.

Kojima immediately hung up.

"What an ass," Newcomer muttered under his breath. *Why did I ever challenge him in the first place? I've totally screwed up the case again. Now he's not going to come to L.A. He's mad because I busted him on his bullshit.*

Newcomer knew perfectly well that Kojima liked to be in control and always played games. This would be the third time the case had been blown to smithereens. There would be no more chances.

His Skype line suddenly rang, and Newcomer lunged for it. With any luck, Kojima might be ready to forgive him. "Hello?" he answered, a hopeful note clinging to the edge of his voice.

"You can hear me?" Kojima asked, sounding like a ghost in the distance.

"Yeah, I hear you fine now. Is everything okay?" *Please say yes,* Newcomer prayed.

"I think so. If I have time I'm calling you, okay? That's it," Kojima simply replied.

"You'll call when you come to L.A.?" Newcomer asked, knowing he had to do whatever it took to repair the relationship.

"Yeah."

"Well, I hope so, because I owe you a lot of money." There was no way Kojima was going to pass that up, was he?

"Uh-huh. Okay. Anyway, thank you now," Kojima said noncommittally.

Newcomer was at a loss as to what else to say. "Anyway, everything's all right?"

The response was that sound of a bubble popping, as though someone was drowning, as Kojima hung up.

"Weird," Newcomer mumbled. He knew that he had to take action.

There were only two possible ways to lure Kojima to L.A., and so far money hadn't proved a strong enough incentive. He was running out of tricks. That left just one thing to try. He quickly composed an e-mail.

> *I'm sorry. I didn't mean anything bad. I just hope you're really coming. I'm not ignoring you. I'm really stuck in San Diego and am using a friend's computer. I really miss you and want you to come to LA. Besides, I owe you lots of money. Come soon so we can see each other in person! Your special friend, Ted.*

It was Ken Kojima who responded.

> *Hi Ted. Yoshi left this morning and will arrive in LA on Monday afternoon. He thinks he can see you and get money because he's not bringing so much.*

That was the answer he'd hoped for. All that mattered was that Kojima was on his way.

Newcomer promptly took the next step. He contacted U.S. Immigration and Customs Enforcement (ICE) officials. Would they please make sure an alert had been placed on Kojima? He also wanted to know the moment that Kojima landed in L.A. Then he wrote Ken an e-mail.

> *Thanks for your message. I'm leaving San Diego now and will be home in two hours. I'll keep my cell phone on and will have cash ready for when Yoshi contacts me. I also have his antiques.*

That was it. He'd promised him money, antiques, and sex. What else could Kojima possibly want?

Monday came and went without word from Kojima. Newcomer picked up the phone and gave ICE agent Jamie Holt a call.

"Hey, I've got an alert on this guy Kojima. He was scheduled to arrive from Japan yesterday, but I haven't heard a thing."

Holt checked it out. "We got nothing. Sorry, but there's no Yoshi Kojima showing up on any Customs entry records for LAX yesterday."

Newcomer began to freak. How in the hell had Kojima snuck into the country? Then it hit him.

Son of a bitch! Yoshi must be traveling on his American passport!

There was no way to verify a damn thing. All he could do was sit and wait for Kojima to call and say, *I'm here in L.A. Meet me.*

In the meantime, previously ordered butterflies kept arriving for Ted Nelson. The packages, sent by his son, slipped past FWS and Customs as easily as Kojima seemingly had. Ken always followed up with an e-mail to confirm that Ted had received them. He quickly became Newcomer's lifeline to Kojima, and Newcomer was determined to keep the connection open.

> *Hi Ken. I received the package of beautiful Ornithoptera Victoria. Thank you. Please tell Yoshi to call me when he can. I really want to see him. I am waiting to hear from him and want to spend lots of time with him. I will change my schedule so I can be with him.*

He'd done everything he could to hear from the man, but Kojima remained elusive. At this point Newcomer started to lose it. He had the indictment and the warrant, only Kojima was nowhere to be found.

He kept his cell phone on 24/7, afraid that he might miss a call. Stress headaches set in as the pressure continued to build and little things in his personal life began to go awry. He'd always been a stickler for orderliness, down to folding his underwear and socks with military precision. Now his laundry was a jumbled mess, he couldn't find his watch, and his head felt thick and muddy.

He'd expect to feel this way about losing his wife, his cat, his best friend, but not about Kojima. What in the hell was happening to him? He sent a desperate e-mail addressed to both father and son.

> *Hi Yoshi and Ken. I hope Yoshi's trip is going well. Yoshi please call me. I miss you and really want to talk. The O. Victoriae are beautiful. Already sold for good profit. O. goliath should be delivered tomorrow. Please call. I want to talk and see you soon.*
> *Your special friend*

Newcomer had never felt like more of an outcast.

At the end of five days he finally received a glimmer of hope.

> *Hi Ted. Yoshi is in Utah near Nevada border. He's staying at his friend's house. His phone is working. You don't know his number? Call him. I leave for US soon and he'll pick me up. Meanwhile, I must take care of his job.*
>
> *Ken*

Why hadn't he thought of that before? Newcomer quickly punched in the number. He might as well have tried to reach the Queen of England for all the good that it did. There was no answer as a game of cat and mouse ensued.

Well, that was a great suggestion. Now Yoshi can reject both my e-mails and my phone calls. This is really beginning to suck big time.

Newcomer was on the road when his cell phone rang unexpectedly. It had to be Kojima. Instead, it was his confidential informant.

"Hey, you're not going to believe this, but I just got a call from Kojima. He's in Arizona with a friend collecting butterflies."

That confirmed it. Kojima really had slipped into the country. Even worse, he was contacting everyone but Ted Nelson.

"Okay, let me know if you hear from him again," Newcomer said with a resigned sigh. He pulled into a McDonald's parking lot. He might as well drown his sorrow in a *Super Size Me* Coke. He was on his way in when his cell phone rang again.

"Hi, Ted. This is Yoshi."

"Yoshi! How are you? Is everything all right?" The words tumbled out, attempting to keep pace with his heart. Newcomer had never been happier to hear from anyone in his life.

"Yeah, yeah. Everything fine. I'm driving to the north rim of Grand Canyon with my friend. We collecting *kaibabensis* larvae here every day."

"That's terrific!" Newcomer replied, his hopes revived.

"Yeah, it's very nice." Kojima described his location down to the road that they were traveling on. "We're going to Las Vegas tonight. I wish you were here."

"I can't wait until you come to L.A. We're going to spend lots of time together, and I have all your cash for you," Newcomer eagerly said.

"Okay. I see you soon," Kojima replied. "I fly to San Francisco in a few days to meet my son. Then we'll come to L.A."

"Great, Yoshi. I promise we'll have fun," Newcomer vowed.

He hung up and immediately called the Grand Canyon park rangers' office. "Listen, I don't know how Kojima got there, but I have an arrest warrant for him. Do you think you can track him down?"

Newcomer would have sent a SWAT team if possible. He no longer cared who snagged Kojima, as long as he was caught. Otherwise, Kojima could just as easily disappear into thin air again.

Newcomer didn't stop there but called the FWS agents in Las Vegas, not wanting to take any more chances. "Be on the lookout for Kojima," he warned.

They all set off on a wild-goose chase, without any luck.

I can't believe it. First he duped Mendoza, and now it's happening to me, Newcomer fretted. *I'm about to be outfoxed.*

There were no more calls from either his confidential informant or Kojima. All communication abruptly dried up.

My entire life is falling apart, Newcomer brooded. Then he received one last e-mail from Ken.

> *Hi Ted.*
> *Did you talk to Yoshi yet? I leave tonight for San Francisco. He's still in Utah for a few days and will contact you. I've also attached a photo of a rare butterfly that no one has caught for 30 years and people can no longer get. A lot of collectors want it. The price is $8,000 for the pair. Please advise.*

Like father like son. They'd stay in touch as long as Newcomer was willing to shell out money. Still, it was a good sign. Newcomer was even more surprised to receive an e-mail from Kojima that same day.

> *Hi Ted. This is Yoshi. I am going to pick up my son tomorrow and transfer to LAX but I have no hotel yet so after check in I advise you. If not busy I will see you. My son wants to see you. Yoshi*

Newcomer couldn't have received better news.

Okay, today is Friday July 28. I'll give them the weekend. The countdown had officially begun.

Three days later, Newcomer was nearly jumping out of his skin. He had yet to hear a word from them. *They should have arrived in L.A. by now,* he thought, as his worry level began to shoot off the Richter scale. Could this be an elaborate hoax being played by Kojima and his son? Where could they possibly be?

Newcomer received his answer that afternoon when ICE agent Jamie Holt called. "Hey, your guy Yoshi just popped up on a manifest out of Japan. He's inbound and due to land at LAX in thirty minutes," he told him.

Newcomer's world came to a standstill.

"That isn't possible. Kojima's already in the country," Newcomer replied, momentarily confused. There had to be a mistake. Perhaps Ken Kojima hadn't flown to San Francisco after all but was arriving at LAX.

"Nope. The manifest says Hisayoshi Kojima. That's your guy, right?" Holt double-checked.

"Damn! Don't let him go, whatever happens. I'll be right there!" Newcomer flew into emergency mode as he jumped into his vehicle, flicked on the lights and siren, and raced as fast as he possibly could to the airport.

A million thoughts sped through his mind during that time. Son of a bitch! If it was true, if he really was just flying in from Japan, then

Kojima was one paranoid genius. He'd been smart enough to create a fake itinerary and tell everyone, figuring that word would get around. That way if Fish and Wildlife agents were looking for him, they'd also be fooled.

Newcomer ditched his car and dashed into the Customs area, where Agent Holt was already waiting for him.

"Is he here yet?" Newcomer asked anxiously.

"His plane just landed, but there's nothing to worry about," Holt assured him.

The felony warrant for Kojima was already in the FBI's National Crime Information Center, a computerized index of criminal-justice information. An alert for his arrest would pop up the moment that Kojima tried to clear Immigration.

"Great. Listen, I don't want him to see me. I'm going to let another agent make the arrest and initially question him. Kojima can't learn my identity yet," Newcomer explained.

"No problem. Just plant yourself somewhere in the area and watch the action," Holt instructed.

"I'll be right here," Newcomer said, and excitedly took cover behind a large pillar. Everything that he'd worked so long for was about to come to fruition.

Maybe it was his nerves, but the wait seemed to take forever as a crowd of people streamed by. *I wonder if this is what Times Square is like in the middle of rush hour,* he mused when Holt suddenly appeared by his side.

"Hey, what does your guy look like again?"

Newcomer's body went numb, as though he'd been dunked in ice water. "What do you mean?" he asked, his voice sounding foreign to his own ears.

"Well, there are only about five people left, and no one in Immigration has seen Kojima," Holt reluctantly admitted.

This couldn't be happening. It simply wasn't possible. Newcomer scanned all the faces flying past as a layer of flop sweat now clung to him like a second skin. "He was on the flight, right?"

"Yeah, yeah. He was definitely on it. Just hold on a moment. I'll be right back."

Newcomer's pulse raced so fast that he wouldn't have been surprised if the room had started to spin. He took a deep breath as Holt hurried toward Immigration. Kojima could still be somewhere in the area.

Where the hell are you, Yoshi? Kojima's a master at deception, and I'm still his damn apprentice.

Holt appeared once more by his side, her face now white as a ghost. "He was here all right, but he somehow got through."

"What do you mean, he somehow freaking got through?" Newcomer demanded in disbelief.

Holt didn't have to say a word. Newcomer already knew there'd been a screwup that would be immediately buried.

Newcomer's blood pressure shot through the roof as he quickly began to scan a sea of Asian faces. Four hundred people had disembarked from the Japan Airlines flight, and in the panic of the moment, every face looked exactly alike to him. Kojima had done it again. He'd made fools of them all. Then Newcomer caught sight of an oddly familiar figure scurrying down the hallway.

He was dressed in shorts, a corduroy blazer, white socks, tennis shoes, a baseball cap, and his ever-present fanny pack while rolling a large suitcase. He'd obviously been cleared by Immigration and retrieved his luggage. Kojima now handed his declaration form to a Customs inspector, who blithely waved him through. For chrissakes, Kojima was heading for the exit!

Newcomer sprang into action and rushed up to the nearest uniformed Customs inspector. "I'm working undercover and have a warrant for that guy in the brown blazer with the black bag over there. Can you pull him aside?" he asked, flashing his FWS credentials.

The Customs inspector raced over and stopped Kojima just as he was about to walk out the door. "Sir, you need to step over here," she instructed.

"Why? Is there a problem?" Kojima questioned.

"We just need to check something," she said as Newcomer caught Holt's attention and silently pointed to Kojima. The ICE agent promptly took over.

"You're under arrest," she informed Kojima, and slapped a pair of handcuffs on him.

Newcomer watched as Kojima was led to an interrogation room, not taking his eyes off him until he disappeared from sight. He'd deal with him tomorrow. There were other things still to be done tonight.

He turned to leave as a Customs and Border Protection supervisor pushed his way through the crowd and planted himself in front of Newcomer. "Who the hell are you, and what do you think you're doing in my Customs area?" he demanded.

This was the last thing he needed to deal with, a supervisor who felt threatened and wanted to mark his territory. He'd gotten his knickers all in a twist because Newcomer commandeered one of his inspectors. The man didn't appear to be the least bit interested in the arrest of a felon. It took all of Newcomer's self-control not to turn and bite his head off. He had no intention of apologizing or asking for permission. Newcomer left the supervisor standing where he was and moved on to the next task of the night.

If I know Yoshi, the first person he'll call is his father-in-law, Charles Hanson. Newcomer needed to gather what information he could before that channel was closed down. It was a toss-up as to where Hanson would be found, either Southern California or the mountains.

It's the end of July. My money is that he's not baking in the desert but cooling his heels in the mountains. Newcomer headed there now. It was a full hour drive from the airport on a good day, and it's never a good day in L.A. Newcomer was immediately swallowed up in traffic. *Damn, I've got to get to the man.* Time was of the essence, traffic was crawling, and he had to act fast.

He had only one option. Newcomer called the local state Fish and Game agent and explained the situation. "Will you do me a favor and go talk to this guy? I have a list of questions that I'd like him to answer."

"Can do," the agent agreed, and headed over with his partner. It was 7:40 PM when they arrived at Hanson's door. They introduced themselves and asked if Mr. Hanson would mind talking to them.

"Not at all. What's the problem?" Hanson pleasantly inquired.

"We have a few questions regarding Yoshi," the agent said.

The conversation rapidly deteriorated at that point. Hanson professed to be shocked to learn that Kojima was under investigation and insisted he knew nothing about butterflies. In fact, he was outraged to even be asked such a question. He'd met Kojima thirty years ago in Japan. They'd become good friends and shared a common interest in antiques.

"I see. So your relationship is—" the agent began, only to be sharply cut off.

"None other than the fact that he lived in L.A. for a couple of years and then went back to Japan. We e-mail each other once in a while to say hello, and he comes here every so often." Hanson quickly corrected himself. "Not *here* but to the United States. That's all I know. Now I'm going inside."

"Just a second," the agent intervened. "Have you ever associated with anyone dealing with butterflies in this state?"

Hanson visibly stiffened at the question. "I have no idea what you're talking about," he replied angrily. Each word flew from his mouth with the power of a bullet. He couldn't believe the nerve of these agents. How dare they come to his house and question him? "We have so much crap going on in the world, and you're worrying about *butterflies*? Go on, get out of here. Go, go, go, go, go, and leave me alone."

"Just one more question." The agent stopped him from closing the door.

"What more can you possibly want?" Hanson nearly howled.

"Do you have a daughter, and if so, are you Yoshi's father-in-law?"

Hanson stared at the agent as though the man had totally lost his mind. "What? I can't believe this! No, I don't have a daughter, and no, I am not Yoshi's father-in-law. Don't come back here again!" he raged, and this time managed to slam the door.

Newcomer listened in stunned silence to the agent's report. If Yoshi wasn't married, then who in the hell was Ken, and why would he claim that Charles Hanson was his father-in-law? Son of a bitch! Had Kojima been lying to me about everything all along? And if Ken wasn't his son, then whom had he been corresponding with? Exactly what was going on?

Newcomer's brain was on overload as Kojima's labyrinth of lies now began to unravel. It felt as though he was on the receiving end of a sucker punch. Even so, he fought traffic back to his office to finish the remaining business of the night. There were still Kojima's belongings to be gone through.

He felt a tinge of excitement as he picked up Kojima's cell phone. Who knew what secrets it held? Newcomer turned it on and scoured the menu for information. Charles Hanson was on his contact list, along with a host of Japanese and overseas numbers. Then he caught sight of one particular listing that sent chills down his spine. It was for National Geographic.

Could it be? Was there a possible connection? Perhaps not everything had been a lie, after all. Kojima was a conundrum wrapped inside a puzzle. The only way Newcomer would ever find out was to press the Call button.

He couldn't help but chuckle as he listened to the voice on the other end. The number turned out to be a recording for National Geographic's subscription line. Kojima was a gutsy chess player who had outsmarted his opponents' every move.

His next chore was to thoroughly search Kojima's luggage. Recalling Kojima's tales of bugs stashed in his suitcase, he instantly got the heebie-jeebies and cautiously removed each rumpled item of clothing, afraid of what he might find. However, rather than bugs, bottles of insulin had been carefully rolled up in his shirts and shorts. He'd forgotten that Kojima was diabetic and promptly arranged for his medication to be sent to the Bureau of Prisons.

Yet another surprise awaited him, at the very bottom of the bag. It was the last thing Newcomer had expected to find: a box of unprotected butterflies. They were marked as a gift for Ted Nelson. The butterflies were perfectly legal, yet Kojima had chosen not to declare them. Instead, he'd snuck them into the country, preferring to be a smuggler to the very end.

Newcomer sat down at his computer to perform one last task for the night. He sent an e-mail to Kojima's "son," Ken.

Something happened to your dad in LA. Call or e-mail me right away. I will help if I can.

He pressed the Send button as the truth came crashing down on him. The handwriting on the packages sent by Ken had matched Kojima's handwriting, and all of Newcomer's communication with Yoshi's "son" had taken place via Kojima's e-mail address.

Newcomer's head began to spin as the realization continued to sink in.

Ken Kojima wasn't part of any conspiracy to smuggle butterflies, and he couldn't have been engaged with his father in illicit sexual activity. Rather, it had been a brilliant ruse concocted by Kojima for one single purpose: to draw attention away from himself.

Kojima's son didn't exist at all.

"Ken Kojima" was never heard from again.

Endgame

My obsessions used to be my protections, but now they have taken me prisoner.
—Mason Cooley

It was the first time in months that Newcomer awoke without having to pretend to be Ted Nelson the butterfly dealer. There were no more Latin names to remember, prices to squabble over, or sex talk to contend with. He'd caught his prize last night. It was now time to start pinning his specimen.

Kojima's face lit up when Newcomer arrived at the jail the following morning. His friend had come to bail him out. Then he caught sight of the holster and badge and reality slowly sank in. He looked up at Ted Nelson again, and his smile quickly faded. This man wore a suit and tie, and his face had somehow changed. The friendly boy that Kojima had known was gone. This was a Ted Nelson that he'd never met before.

Newcomer felt nothing more than that he'd caught his prey and a loophole in the illegal trade was about to be closed. Neither said a word as Newcomer led Kojima from the Bureau of Prisons to the U.S. Marshals Service lockup. Then the man that Kojima had traveled 5,478 miles to visit handcuffed his wrist to a metal table. It was Kojima who finally broke the silence.

"Have you been a Fish and Wild guy from the first day we met?" he sadly asked.

"Yep," Newcomer shortly replied, and continued to fill out his paperwork.

Kojima looked away and contemplated that for a while. When he glanced back his eyes landed on Newcomer's wedding ring. "And you're *married*?" he asked in amazement.

"I'm married," Newcomer replied firmly without looking up.

Both men knew it was the final act of betrayal.

Newcomer stood and left as soon as his paperwork was complete. A few more hours, and he'd be rid of Kojima forever. He could hardly wait. He'd had it with the lies, the manipulations, and Kojima's endless games. His last task was to return and escort Kojima to the courthouse for his arraignment. After that, Kojima would be the responsibility of the U.S. Marshals Service until sentencing. It couldn't come soon enough as far as Newcomer was concerned. He had caught, pinned, and mounted the world's most notorious butterfly smuggler. It would send a message out to the world: Don't screw with what's protected and fragile in nature, or you'll be held accountable.

Newcomer returned at 12:30 sharp to escort Kojima to his court arraignment. A deputy marshal brought him from his cell. Kojima's complexion was pasty, and beads of sweat clung to his upper lip like dewdrops. His wrists were cuffed behind his back as he stumbled and fell against the wall.

"Yoshi, are you okay?" Newcomer asked. He was alarmed by his appearance.

Kojima's eyes were glassy. When he spoke his voice was trembling and weak. "I didn't get my insulin today," he muttered, continuing to hug the wall.

"What do you mean? Didn't they give you lunch?" Newcomer asked.

The deputy marshal had yet to say a word. Something was definitely wrong. "Yes, I ate. That's why my blood sugar is spiking," Kojima managed to mumble.

At least now Newcomer knew the problem. It was one that could be easily fixed. He turned to the four deputy marshals who stood watching as though it were a reality show.

"Hey, guys, my prisoner is diabetic. He needs his insulin," Newcomer informed them.

The deputies responded in unison as though they were a Greek chorus. "We don't give medication at the marshals' lockup."

That might be true, but it didn't help solve the problem. This wasn't a stomachache Kojima was experiencing. He was on the fast track to a full-blown insulin attack and was a prisoner at their lockup. "His insulin was turned over to the Bureau of Prisons. Would you please give them a call?" Newcomer asked.

"Yeah, sure, but we don't give medication here," one of the deputies repeated.

It was yet another showdown between employees of two federal agencies. The power struggles seemed endless. To Newcomer's mind, these were his fellow law-enforcement officers. That made them compatriots-in-arms, *muchachos*, his buddies. They were all in this crazy world of the legal system together. Wasn't it an unwritten rule that they help one another out? Surely they'd come to his aid if necessary. "Just call them," Newcomer once more requested.

The deputy stared at him a moment and left. He returned a few minutes later. "Yeah, they have his insulin all right, but they can't take it out of the building."

It was a hell of a start to the day. Newcomer's blood began to boil as the deputy slowly folded his arms. His body language was easy to interpret. *You're on your own, compadre.* This wasn't the way things were supposed to work, but then nothing with this case had gone by the books so far.

Newcomer turned to Kojima. "Yoshi, you're scheduled for one PM, but it will probably be another hour before the judge sees you. Can you make it to your arraignment?"

Kojima simply nodded. His face was now expressionless, and his mouth hung loosely open. He was clearly deteriorating fast.

He's never going to make it. Now what am I supposed to do?

Newcomer caught the four deputies smirking at his dilemma. He knew what they were thinking. *He's the big-shot special agent. Let him figure it out.* He could have easily blown them all away if it wasn't for the fact that he was a nonviolent man.

He took hold of Kojima and sat him on a bench, then focused his attention on the deputies. "Okay, then, why don't you call the federal occupational-health nurse and have her come down? At least she can assess the situation," he suggested.

Kojima was slumping and unresponsive by the time the nurse appeared.

"He's definitely not doing well and needs his insulin," she agreed. There was nothing more she could do as Newcomer now found himself sinking deeper into a bureaucratic nightmare.

The deputies adamantly refused to help until Kojima was remanded to their custody, and that wouldn't happen until after his initial court appearance. Newcomer decided the situation had gone far enough.

"So he's my responsibility, right?" Newcomer asked, already knowing the answer.

"Yep," one of the friendly deputies agreed. "He's not our prisoner until the court turns him over to us."

"Okay, then we're out of here," Newcomer said, and helped Kojima to his feet. They began to walk toward the exit.

The marshals quickly jumped into action. "Hey, wait a minute. You can't do that! He's on the one PM docket. Come back here!" they yelled.

What do you know? Newcomer had finally gotten their attention. "Nope. He's my responsibility and you won't treat him. Call the Bureau of Prisons and tell them we're on our way over," he said, continuing to walk on.

Not one of the deputies made a move to help.

"Oh, yeah, one more thing. You can also tell the judge we're leaving because my prisoner is having a diabetic attack," Newcomer said in a parting shot.

Kojima grunted and leaned against Newcomer a little more. "Yoshi, we're going back to the Metropolitan Detention Center for your medication. I'm sorry, but you're going to have to make your appearance tomorrow," he told him.

Kojima focused on taking one step after another without falling. "Okay," he barely managed to mouth.

"You're going to get into trouble!" one of the deputies yelled after him.

They entered the bowels of the building and began to walk down the long hallway as the cry followed like an angry wraith. Their footsteps echoed dully on the cement, and the overhead light cast a sickly pallor on the white concrete walls. The Bureau of Prisons loomed forebodingly ahead, its underground entrance blocked by a giant steel door.

They were three-quarters of the way when Kojima's legs suddenly buckled and he began to fall. Newcomer grabbed hold and leaned him against the wall, but Kojima's feet flew out from under him.

"I need help!" Newcomer screamed to two security guards.

A chair was brought over and the occupational-health nurse came back down. She took Kojima's blood pressure and heart rate again. "He's not doing well," she said.

Kojima was placed in a wheelchair, and Newcomer rolled him to the Bureau of Prison's doorway, where a twenty-two-year-old guard sat reading a book.

"I'm a special agent with U.S. Fish and Wildlife. My prisoner is having a diabetic attack. We need to get him inside for an insulin shot," Newcomer carefully explained.

The officer barely deigned to glance up. "Well, you can't do that because he's due in court. You'll have to wait until five PM."

Newcomer was tempted to rip the book from his hands. "No, you don't understand. His insulin is inside and my guy's out here," he replied between clenched teeth.

"It doesn't matter. Doesn't he have his initial court appearance today?" the guard countered.

"Yeah, but we're going to miss it." The words rat-a-tat-tatted like a round of ammunition.

The guard never got off his rear but nodded toward a phone on the wall. "Well, you can call over there and see if they'll come down."

Newcomer picked up the phone and reached a Bureau of Prisons physician's assistant. "This is Ed Newcomer, and my prisoner's having a diabetic attack. I've got him out front. He's in a wheelchair, and the occupational-health nurse is with us. We need to bring him inside to get his insulin, or he's going to miss his court date today."

It was as if every employee had been programmed to spout the same response. They didn't take prisoners in the middle of the day. Newcomer would have to return him to the Marshals' Service, and they could bring Kojima back at five. Only then would he get his insulin shot. Were they all robots?

"Well, we're here now. Either he goes to the hospital or you let him inside. What are you going to do about it?" Newcomer growled.

"I'll check with the doctor," the physician's assistant replied in an attempt to get off the phone.

Newcomer and the nurse waited a good twenty-five minutes until he couldn't stand it any longer. "Hey, buddy, can you help us out? This guy is going into a diabetic seizure, and we need to get in there," he appealed to the guard once more.

The guard briefly glanced up again. "Afraid I can't do that," he said, and went back to reading his book.

We've been jerked around long enough, Newcomer decided. It was time to take control. "Here's the deal," he said to the nurse. "I don't know anything about diabetes. So unless he goes unconscious, I have no clue as to what's going on. You're the nurse. You tell me when it's time to call nine-one-one, and I'll do it."

She nodded in agreement. "I think he's okay for the moment. He's still conscious but quickly getting worse."

That wasn't good enough. Prisoner or not, this was no way to treat a sick man. Newcomer walked over and picked up the phone once more. "You want to tell me what's going on?" he demanded.

"We're working on it," the physician's assistant lackadaisically replied.

"Okay, here's the deal. Either you come down here with the insulin in five minutes or I'm calling nine-one-one. Then you can personally explain to the fire department why they had to come to the Bureau of Prisons to rescue a federal prisoner," Newcomer threatened, and slammed down the phone.

He waited a few more minutes. Enough was enough. Newcomer picked up his cell phone and began to dial 911 just as the prison door flew open. A doctor emerged with a smile and a syringe in hand as if everything was fine and he was happy to be of help.

"I'll be back in the morning to pick you up," Newcomer told Kojima before turning to the guards. "Kojima better be on the calendar bright and early tomorrow morning to see the judge."

From there, Newcomer walked over to assistant district attorney Joe Johns's office and explained why Kojima hadn't made it to his arraignment. He was so steamed that he couldn't sleep. Instead he stayed up and wrote letters about the incident to his senators and congressional representatives. He regrets to this day that he never mailed them.

The next morning Kojima was assigned a public defender and pleaded not guilty to seventeen charges related to the sale and smuggling of endangered butterflies. Neither his "father-in-law," "ex-wife," or "son" showed up for him in court. Newcomer pondered what the true story was. Would the real Kojima please step forward? In that respect, Kojima's life was almost like a game show.

Jill Ginstling, his attorney, immediately had Kojima's financial affidavit placed under seal. It was a wise move on her part. The action prevented Newcomer from discovering any of Kojima's other bank accounts. Newcomer had hoped to track down thirty thousand dollars from Kojima's recent sale of *homerus* and *alexandrae* butterflies to a couple of California buyers. It was either another of Kojima's lies, and they'd never sold, or the money was safely stashed away somewhere. Neither the butterflies nor the money was ever found.

Newcomer received his last Express Mail envelope from Japan during Kojima's first day in court. The package contained a single CITES II *paradisea*. It now brought the worth of the butterflies that he'd bought and been offered to over $294,000.

Ginstling met with Joe Johns shortly after Kojima's arrest. She had one burning question. "Just who is this Agent Newcomer, and is he incredibly good-looking?"

Johns looked at her in amusement. "Why do you ask?"

"My client said he was so beautiful that he just couldn't resist him," she replied.

It would have made for a unique defense.

"It's a funny thing. Yoshi wasn't thinking about going to jail when he was arrested. Instead he was thinking about this guy that he loved who ended up stabbing him in the back," Newcomer reminisced.

Joe Johns agreed. Not only had a cagey swindler been caught, he'd ultimately been betrayed by his heart.

Kojima remained as paranoid as ever. He suspected Ginstling of fraternizing with the enemy and fired her within the first few days. She was replaced with Norman Sasamori, a general practice attorney that Kojima found through the Japanese embassy.

Sasamori is best known as the general counsel of the Hiroshima and Nagasaki Peace Project founded by his mother. Shigeko Sasamori had been a thirteen-year-old schoolgirl in Hiroshima when the first nuclear bomb exploded and instantly killed her companion. A quarter of Sasamori's body was severely burned by radiation, and her fingers were scorched to the bone. In 1955 she became a "Hiroshima Maiden" and was brought to the United States by American author and diplomat Norman Cousins. Sasamori underwent thirty reconstructive surgeries. She eventually settled in the United States, married, and had a son who'd now been hired to defend Yoshi Kojima. That would prove to be a strategic mistake on Kojima's part.

Public defenders are always in trial and have vast experience when it comes to dealing with federal prosecutors. Ginstling knew what weaknesses, if any, Joe Johns had. Sasamori didn't have that advantage as 1,116 pages of discovery, along with dozens of hours of video and audio material, were now transferred to his office. "Kojima's interests would have been better represented by Jill Ginstling," Johns stated diplomatically for the record.

A status conference was set for January 2007 for which Johns prepared a plea agreement. It included a sentencing guideline of Level 15. Sentencing guidelines are the formula used by the federal justice system to determine the amount of time a prisoner should serve. Level 15 fell between sixteen and twenty-four months imprisonment. The plea agreement was automatically sent to Kojima's defense attorney.

Sasamori showed up fully determined to face off with Ed Newcomer and Joe Johns. Producing a letter, he placed it in front of them as if it was Kojima's golden Get Out of Jail Free card. Kojima's good friend and "father-in-law," Charles Hanson, had written a letter on Kojima's behalf.

> *I have had the pleasure of knowing "Yoshi" for over 30 years, ever since I was engaged in business in Japan and working with in regards to collecting antique furniture and decorative items. The business relationship developed into a long, lasting friendship that still exists today. He is devoted to family and especially to caring for his aging*

mother despite his own illnesses. I have always known Yoshi to be a hard working, diligent, and honest person.

Johns read the note and was thoroughly unimpressed. "Charles Hanson? Isn't he the gentleman that we talked about indicting for co-conspiracy?" he asked Newcomer.

"That's him," Newcomer replied.

Johns handed the letter back to Sasamori. "We have evidence that Mr. Hanson was a co-conspirator on at least one deal involving an illegal butterfly. We've decided not to indict him at this time."

That was as far as Sasamori's character references for Kojima went.

Johns patiently waited as Kojima's attorney now tried a different tack. Sasamori maintained that neither Newcomer nor Johns could prove that his client was guilty. After all, it was only alleged that Kojima had said certain things.

Newcomer and Johns looked at each other, unable to believe what they had just heard.

"Actually we not only can but will prove it five different ways," Johns shot back, then threw the ball to Newcomer. "Tell Mr. Sasamori what Skype tape Kojima's first statement is on, along with its date and the other separate undercover recordings that we have."

"It's all right here," Newcomer complied. Reaching over, he opened Sasamori's binder with the Skype DVDs and discovery papers inside. He pointed to the index page. "Kojima mentions selling endangered species on disc number twenty-two."

Sasamori looked at both men in what seemed to be undisguised surprise.

"I could tell from his face he had no idea that we had undercover videotapes and recorded conversations," Johns stated. "All I can guess is that he came in to negotiate a plea agreement and hadn't yet gone through the binder."

Sasamori maintained that he still thought fifteen years was a lot of time to ask for what Kojima was accused of doing. Newcomer and Johns swapped a sidelong glance and sat for another moment in stunned silence. Sasamori had apparently also misinterpreted the sentencing guidelines.

Joe Johns played it cool. "Well, you know, it's a very serious crime, but perhaps you misunderstood what Level Fifteen stands for. It refers to the sentencing guidelines, not the number of years that we're asking he serve."

There was no question that Sasamori had been outgunned and outmatched by Joe Johns.

"Sasamori had a lack of familiarity with the federal criminal-justice system. It's a different ball game than doing state criminal-defense work," Johns explained.

Johns had an open-and-shut case. Kojima wasn't about to go to trial with what was on the Skype tapes. He'd been caught in living color making statements and admissions, claiming knowledge of smuggling, and offering to sell illegal butterflies, to say nothing of all the sex talk.

"I would have slaughtered him if he'd tried to weasel his way out of it," Johns firmly attested.

Sasamori made a final gesture to try to help his client. What if Kojima agreed to tell what he knew about the butterfly trade and cooperate with the authorities?

Newcomer turned him down flat. "Yoshi had consistently lied to me. I didn't trust him. He'd do it again or give me some chicken-shit information and get credit for cooperating."

Newcomer believed he'd caught Mr. Big, the man who got butterflies that no one else could get. Take him out and it would shut off the wires.

"I'm done dicking around with Yoshi, no pun intended. I want him to go down hard," Newcomer told Johns.

Johns responded by asking that Kojima be sentenced to twenty-four months, the maximum time allowed under the sentencing guidelines. Sasamori countered with fifteen, while the Probation Department recommended that Kojima serve eighteen months in prison.

Judge George Schiavelli had final say on April 16, 2007. "The species here are too critical. Their value is too critical. Obviously this is a person who has a big impact on wildlife crime. So I'm going to sentence Mr. Kojima to serve more time than what the Probation Department recommends."

Kojima was sentenced to spend twenty-one months in prison. In addition, he was ordered to pay a $30,000 fine, $7,656 in restitution to the USFWS, and $1,175 court assessment. None of the money was ever collected.

Kojima served the remainder of his sentence at California City Prison located off Twenty Mule Team Road in the desolate Mojave Desert. The town is an amalgamation of tired houses and worn-out buildings dotted with dust and tumbleweeds that meander down the road. It could easily pass as a set for an old episode of *The Twilight Zone*.

The prison rises from the desert like a monolithic mirage a few miles out of town. Its towering backdrop of brown mountains bears the scars of crime victims, their perpetrators off-road vehicles. A prison for deportable aliens, it's described as a medium-security facility, though the façade suggests otherwise. The building is encircled by three rows of ten-foot-tall cyclone fence with layers of concertina wire on the bottom and top. The prison houses predominantly Mexican criminals and is known for its warring drug gangs.

Busloads of people pay homage near the site on the thirteenth of every month. They come not for the prisoners but to pray at a humble shrine called Our Lady of the Rock. There they stand in the frigid cold and under the burning-hot sun with rosaries, religious pictures, and cameras in hand. It's on that day that the Virgin Mary is said to appear in the sky, though only true believers can see her. Kojima never once spotted her from where he was locked up.

Meanwhile, his story made big news back in Japan. WORLD'S MOST WANTED BUTTERFLY SMUGGLER ARRESTED AT LOS ANGELES AIRPORT, blared *Japan Today*. The article spawned a slew of differing opinions on the newspaper's Web site. "This is a much larger issue involving Yakuza, bribes to authorities, Chinese triads, and a mysterious Czech woman," one reader posted. Another asserted that it wasn't a Czech woman but rather a Russian transvestite who was involved. Still someone else believed Kojima had resorted to a life of crime to support his sex fetish. They insisted he had an addiction to costumes and maid cafés, featuring young girls dressed as naughty maids who wait on customers hand and foot.

Only one person blamed the Japanese government for having looked the other way.

The news had an impact in the United States as well. Some collectors had actually believed Kojima was an undercover agent running a sting operation. How else could he have gotten away with selling illegal butterflies for so long? Others were angry with him for having fouled the waters for legitimate interests, while there were those who felt he'd been entrapped even though he had known the law.

"Yoshi reminds me of the con man in the movie *Catch Me If You Can*. He was that good until he got caught," stated lepidopterist Bob Duff.

Even the CI had second thoughts. "I have trouble putting anybody in prison for catching a bug." Would he have turned Kojima in if he had realized the consequences? "I'd have to search my conscience about that. I wanted Kojima to quit smuggling and dealing illegally, but I don't know that it was worth jail time."

Newcomer had hoped the Japanese police would cooperate after Kojima's arrest. He made numerous requests to obtain Kojima's computer hard drive with its files of all his suppliers and customers. It was the only way to finally bring Kojima's network down.

Newcomer was warned that getting anything from the Japanese

police would be a long shot. The information proved correct. They not only regard undercover work as undignified but also consider those Japanese officers who do it to be dishonorable. The fact that an American undercover agent caught a Japanese national in a sexual web only made the matter that much worse.

Investigative leads continue to be forwarded to the Japanese police in the hope of identifying U.S. customers. To date, there has been no response.

WITH KOJIMA IN PRISON, Newcomer turned his attention to finishing up the roller-pigeon case. By May 2007 he was ready to take down his targets. He had everything he needed on surveillance and audiotapes, along with hard-core evidence of dead hawks that had been collected. Seven suspects in California were to be served with warrants and arrested. The plan was to take them down all at once to send a message to the rest of the pigeon breeders throughout the country.

Two dozen law-enforcement officials gathered the day before and listened as Newcomer worked out the details. Then they spread out across Southern California and hit hard the following morning. LAPD's Major Crimes Division flew helicopters over some of the rougher neighborhoods as Newcomer conducted his arrests.

The last stop was the home of Juan Navarro, president of the National Birmingham Roller Club. Newcomer searched his garbage and found a dead hawk in it that very day. By the time the raid was over, word raced across the Internet like wildfire, creating enough paranoia to temporarily stop the killing of hawks.

Most of the breeders arrested lied about their crimes. They denied having ever killed a bird. Newcomer was as methodical and precise as ever. He put together two notebooks containing every damning scrap of evidence against the men that he'd collected.

Their defense attorneys went to see Joe Johns, certain their job would be a cakewalk and the charges easily dropped. A bombshell hit when Johns once again presented notebooks filled with incriminating evidence.

"Holy shit. Who put these together?" one defense attorney asked after quickly looking through the organized and indexed binders.

Johns allowed a smile to creep across his face. "The agent for this case was once a lawyer. Special Agent Ed Newcomer did it."

All seven defendants pled guilty, knowing that their proverbial goose was cooked.

Under the law, the killing of thousands of hawks each year is considered a Class B misdemeanor. The seven men received suspended sentences of six months, along with five years probation, while a few were required to do community service. Only Juan Navarro was convicted on sixteen Migratory Bird Treaty Act (MBTA) counts and fined twenty-five thousand dollars. All of the breeders were allowed to keep their pigeons.

"You can't deter people with Class B misdemeanors. It's a two-hundred-fifty-dollar fine. You get penalized more for speeding in the HOV lane," Newcomer noted scornfully. "The bottom line is that we can't effectively enforce the laws with the penalties available to us."

Nothing will change until penalties are stiffened and jail time becomes mandatory. Until then, violations will continue and migratory birds will be viewed as fair game.

When it comes to wildlife crime, it's business as usual. The illegal trade continues to be nothing more than one big revolving door.

OPERATION HIGH ROLLER WAS PUT TO BED, and Kojima was released from prison in February 2008. He was immediately taken to LAX and deported. As a convicted felon, he will never be allowed to return to the United States. Kojima had spent thirty years in America, longer than he'd lived in his own country. Even so, that wouldn't stop him from continuing his butterfly business, although he had no idea what might await him when he arrived back home.

Journey to Japan

I do not know whether I was then a man dreaming I was a butterfly, or whether I am now a butterfly dreaming I am a man.

—Chuang-tzu

Kyoto is an ancient city, a maze of quiet alleys with warrens of old wooden houses, vermillion torii gates, and temples with tiled rooftops that gently curve like waves. Geishas dressed in kimonos garnished with elaborate obis flit in and out of doorways. Few signs are in English. It's a city of secrets that keeps strangers at bay.

A young Japanese woman approached a store in central Kyoto whose brown shop curtain was tattered and torn. Even so, she was able to make out the distinctive figures that spelled KOJIMA'S ANTIQUES.

Antique blue-and-white porcelain bowls sparsely lined the window, and the interior of the store was dark and in disarray. A note on the door instructed visitors to ring the buzzer. She did so, and a voice responded over the intercom.

She was told the shop was permanently closed and to just go away.

"Was he there? What did he say?" I asked as she walked back to where I stood.

I'd heard of the case when it broke and had become obsessed with the man. This wasn't the first I'd dealt with a story involving the U.S. Fish and Wildlife Service. I had written about special agents for years, listened to their tales, and even woven their adventures into a mystery series. However, this undercover operation was different. It captured my interest like no other. I wanted to know what it was that made Kojima tick. Why would someone become a butterfly smuggler? However, Kojima had already served his sentence and been released from California City Prison. He'd returned to Japan and refused to speak to anyone.

I'd called Charles Hanson, but the two were no longer on speaking terms. Hanson had lent Kojima money while in prison and never been repaid. He was furious and hadn't been in touch with Kojima since he'd been deported.

I'd met butterfly collectors in California who had known and been friendly with the Japanese dealer. All felt as if they'd been deceived by him. They had believed he worked for National Geographic and had a wife and son, though no one had ever met them. Most spoke off the record, not wanting to be associated with Kojima. They whispered of a partner who dealt in drugs and hinted Kojima wasn't a lone wolf in the business, none of which could be confirmed.

I'd then called Kojima's former lawyer, Norman Sasamori. He refused to answer even the most basic questions. "Maybe you should go talk to Mr. Kojima in Kyoto," Sasamori suggested sarcastically before abruptly hanging up.

Why not? I thought. I took him at his word and flew to Japan on a mission without a backup plan. Kojima could have been anywhere in the world rather than at home. The other quandary was that I didn't speak any Japanese, making it difficult for me to track him down.

I asked everyone I knew for any connections in Japan and was fortunate enough to be introduced to a Japanese woman who agreed to help. Then came the dilemma of arranging the trip. She was busy with her own job in Tokyo and rarely answered my e-mails. I was rebuked as a "pushy Westerner" when I tried to follow up. I began to feel as if I was walking on cultural eggshells.

We arranged to meet at the Tokyo train station, a busy maze that's nearly impossible for a foreigner to navigate. However, I figured that I would easily stand out in the crowd, being the only redheaded Caucasian in the place. A half hour flew by as we frantically exchanged cell-phone calls while trying to locate each other in the station. We eventually wound up standing on opposite sides of the same pole.

A two-and-a-half-hour bullet-train ride was all it took for cosmopolitan Tokyo to fade into memory, replaced by the mystery of Japan's ancient capital. Kyoto is as enigmatic and multilayered as Kojima himself. In Kyoto one worries not about traffic but about being hit by residents furiously pedaling to and from work on their bicycles. The women ride in high heels, and both sexes navigate while holding a cell phone up to their ear.

They keep watch for speeding Mercedeses, Hummers, and Lexuses, the favored vehicles of the *yakuza*. The crime organization rivals the American Mafia, except that their members are officially registered with the police, play an important role in business affairs, and have a large presence in Kyoto.

We checked in to our hotel, then flagged a taxi, whose white-gloved driver took the proffered address. He seemed to have no idea where it was that we wanted to go. We were eventually dropped off on a seemingly endless street with no numbers and began our search for Kojima.

Kojima once had a Web site for his butterfly trade. I'd perused the site before he returned home and shut it down. One page had been dedicated to his antiques shop, with a few of his favorite pieces highlighted. A butterfly dish had caught my eye. I spotted it now in the window of a run-down three-story house. The shredded *noren* above the entrance verified that it was Kojima's place.

I'd spent nine months immersing myself in countless hours of Skype DVDs and audiotapes learning all that I could about Kojima. I'd methodically delved into Ed Newcomer's brain. Finally, I'd traveled halfway around the world to meet the man. Now I feared that I'd receive the same reception as everyone else who had tried to speak with Yoshi Kojima. I refused to be added to a long list of people who had hit a dead end.

It was agreed that Yuko would approach him first as a customer while I remained out of sight. I'd interviewed a number of people who knew Kojima, and it was possible that he'd been tipped off about my trip to Japan. I would join her once she was inside. So much time and planning hinged on just two things: Would Kojima be in Kyoto, and would he let us into his shop?

Kojima was home, but he had no intention of letting Yuko inside. I had traveled all the way to Kyoto for nothing.

We walked down Teramachi Street with its scattering of hidden Buddhist temples and electronics stores that were closed. Stepping into a small bookstore, Yuko asked the owner about the antiques shop across the street. The woman replied that it had been closed for two years. She knew nothing more.

We headed back out, only to spot Kojima sweeping the floor of his shop.

"Hurry! See if he'll let you in," I urged, refusing to believe this was the end of my journey.

But Kojima saw her coming and disappeared as fast as a tarantula into a hole. This time he refused to answer the buzzer.

There had to be a way to figure this out. We trekked down the block to a large shopping arcade and main street with its row of expensive stores—Brooks Brothers, Cartier, Benetton, and Takashimaya. Real estate in the neighborhood had to be worth a fortune. Then we wandered back to Kojima's. The shop's steel shutters were now pulled down even though it was the middle of the afternoon. Kojima couldn't have been any more clear. He wasn't available to anyone. He was as elusive as a shadow.

I drowned my worries in too much sake that night as transvestites sauntered back and forth near his neighborhood. Some worked as shopgirls, the only clue to their gender being the size of their hands. I wondered if one might be the mysterious Russian transvestite that I'd heard about.

A Shinto shrine with its bright lanterns glowing yellow as full harvest moons beckoned us inside. We entered the courtyard, and Yuko demonstrated proper shrine etiquette. I followed her every move. Water flowed into a large stone trough, where we ladled it first over one hand and then the other, after which we rinsed our mouths to purify ourselves. Approaching the shrine, we pulled a long hemp rope to ring a bell in hope of gaining the deity's attention. A lucky five-yen piece was tossed into the offering box, then we made two deep bows, two claps of the hands, and a prayer: *Please let me meet Kojima and be given the chance to speak with him.*

I bowed deeply once more and slowly backed away, careful not to turn too soon and insult the gods by showing my rear end.

Finally, a fortune-telling machine with its animated dragon begged to be fed. The mythic creature danced and shook its head as if laughing at the whimsy of fate. I slipped in a coin, watched the dragon grab a slip of paper in its jaws, and dropped my *omikuji* into the dispenser.

How good your fortune is! Everything will be all right. Nothing to worry. Work hard at anytime. Do not give yourself up to drinking or illicit love.

"That's a good fortune. You must keep it," Yuko said through our sake-induced haze.

I needed all the good luck I could get.

We staked out Kojima's shop early the next morning, its shutters still closed at 8 AM. I decided to wait him out at a café down the block that advertised seasoned cod-roe ice cream. A short while later we heard the angry clank of steel shutters opening, the corrugated armor piercing the air. I jumped up and ran to the street in time to see Kojima riding away on his bicycle. A flock of crows in a nearby tree cackled at my misfortune.

Yuko remained seated as I stood at the shop window and gazed at the antique bowls with their ancient designs. The one that kept catching my eye was a dish in the shape of a butterfly. It matched the blue enamel butterfly pin that I'd purposely worn that day.

I was still studying the dish when my skin prickled and the hairs on the back of my neck bristled like those of a porcupine. I instinctively knew that I was being watched. Spinning around, I found myself face-to-face with Yoshi Kojima. This was it, my one and only shot.

"Good morning. Is this your shop?"

Kojima stood in a faded baseball cap, denim shirt, and khaki shorts with a black fanny pack encircling his waist like a ring on a tree. He had aged, but his skin was still smooth as polished stone. "Yes," he said, eyeing me warily.

"I love that butterfly dish, and the antiques in your window are beautiful. What time do you open?"

"The shop is closed," he replied slowly.

"Just today, or will you be open tomorrow?" I inquired, feigning ignorance.

He repeated what I already knew. "No, the shop is permanently closed."

"That's too bad. I was hoping to buy that dish. I have a special fondness for butterflies, especially blue ones," I said, and pointed to my pin.

I'm not sure what I expected of him, probably someone unfriendly and gruff. I wouldn't have been surprised if he'd refused to speak and closed the door in my face. Instead, he unlocked the shop and motioned me in. I silently thanked the gods at the Shinto shrine for my change of luck.

Kojima rolled his bike next to a second bicycle in his shop and then brought over the butterfly dish.

"I bought these at an auction over twenty years ago. They're part of my personal collection," he said. "I think there are four more plates in the set."

"Are you sure you want to sell them?" I asked.

He removed his cap to reveal a pate of thinning dark hair. "Yes, I no longer care. I have so many things." He drew my attention to three antique screens that decorated the walls of his store.

"These are worth more than a million dollars," he confided.

One was four hundred years old and featured two tigers with eyes of gold. They glared at me ferociously, as if aware of exactly who I was and why I was there.

The tansu chests, bronze pots, and lacquer trays were all beautiful, but the shop itself was a wreck, with sagging wooden beams and a sinking cement floor. The railroad-style house was dirty and reeked of mildew and sewage from a sewer pipe that had backed up under the floor. Kojima was in the midst of a fight with his brother as to who should pay to have it fixed. They had clearly reached a standoff.

Kojima didn't rush me from the shop but now began to talk. He asked where I was from and what I did for a living. I could either come clean, and risk everything, or concoct a story. I morphed into the owner of a small knickknack shop specializing in antique toys. In turn, Kojima related that he'd lived for thirty-eight years in the United States, where he'd had a travel agency in L.A. along with an American wife and son. I felt like a fraud but played along.

"What will you do with all these antiques now that your shop is closed?"

"Oh, my son might take over the store. Everything goes to him," Kojima said. And there was more. "My son works for National Geographic. In fact, he's in South America right now. I also used to work part-time for them. They paid me fifteen thousand dollars a month."

"Really? I know some photographers who shot assignments for the magazine," I replied, and mentioned a few names.

Kojima shook his head. "No, I wasn't a photographer. I did something else."

"What did you do?" I probed.

"I took photographers and film crews into the jungles of Bolivia, Peru, Colombia, and Ecuador," he said.

"To do what?" I asked.

"To find butterflies," he replied with an enigmatic smile. "I'm very famous for them."

"South American butterflies are beautiful. My favorites are morphos," I said, and pointed again to my pin.

"I have lots of them, but that's what got me into trouble," Kojima disclosed without the least bit of prompting.

"What kind of trouble?" I asked, surprised he would reveal so much of himself so quickly.

"For selling butterflies. I was in prison in the U.S. for two years, but I was caught by entrapment." Kojima began to weave his story as deftly as a spider weaves a web.

A young boy had come to his Los Angeles home, told Kojima that he loved butterflies, and wanted Kojima to teach him about them. He was a nice-looking boy, and Kojima had taken him at his word. However, Kojima soon became ill and returned to Japan for bypass surgery. The boy grew furious. He'd wanted to trap Kojima and was angry that he hadn't caught him.

Kojima returned for the L.A. insect fair two years later, and the boy approached him again. This time he asked Kojima to sell him illegal birdwing butterflies. Kojima didn't know that the nice-looking boy secretly had a bad heart. He was really a Fish and Wildlife agent. Kojima sent the butterflies from Japan and was then lured back to California. That's when Ed Newcomer pulled out a gun at the airport and arrested him. What Newcomer didn't know at the time was that Kojima's son had been with him.

"I told my boy to run! He was so upset that he was crying." Kojima grew more animated by the minute. "Fish and Wild went to my ex-wife's house and harassed her. Even the FBI joined the search, but they couldn't find my boy. That's because Ken has a different last name. They tell me, 'When your boy return we catch him.' But I say, 'I have no boy. How come you saying my boy return?'" Kojima laughed.

Had Newcomer been wrong? Did Kojima really have a son? Kojima pulled out his cell phone and produced a photo of a handsome twenty-nine-year-old Eurasian as evidence.

"That's him. That's my boy," he proudly claimed. "He speaks seven or eight languages. He's a very good businessman. After that, Fish and Wild tried to put me in prison for sixty years. You should read articles about me in the *New York Times*. I'm very famous."

"I'll do that," I replied, and then dared to ask the million dollar question. "Can I buy butterflies from you if I want to take them home with me?"

"I can sell them to you, but big trouble if Customs finds them. It's better I send to you by Express Mail. That way no problem."

It was interesting to know that nothing had changed since his release. Kojima was back in business.

"So you still sell to customers in the U.S.?" I asked, growing increasingly comfortable in my new role.

"Yes, but mostly in Japan and Europe," he said. "I have a half million papered butterflies in my house." The unmounted butterflies were stored in boxes protected by mothballs.

I promised to return later that day to buy the dishes, and Kojima agreed to show me some of his insects.

I dashed to the café, gave Yuko the good news, and went to buy a gift for Kojima as a token gesture. Candy was out, so I settled on a box of freshly made rice crackers. Who knew that a number of senior citizens choke to death on them every year in Japan? The next stop was the same Shinto shrine where I had so successfully prayed the night before. I threw in a five-yen piece, thanked the gods, and prayed again.

We made our way back and Yuko waited at the café as I returned to Yoshi's shop and rang the buzzer. Kojima came downstairs and let me in. The butterfly dishes were already neatly wrapped.

"Are you in Kyoto all by yourself?" he asked.

"Yes. It's lonely, since so few people speak English. I'm very happy to have met you," I replied.

"It's smelly in here. Let's go for coffee," he suggested.

Kojima shuffled down the street with his shoulders hunched. He walked like an old man, though he was only fifty-nine years old. He'd recently fallen off his bicycle, nearly been hit by a car, and cracked his tailbone.

We entered the same café where Yuko sat reading a book. *Please don't let her look up and give me away,* I prayed as we walked right past her.

"You are very lucky to catch me. I was just in the hospital for three weeks. I only got out four days ago," Kojima revealed.

I silently thanked the Shinto gods again for my good fortune. "What was the problem?"

"I'm so dizzy all the time. It's my heart. I must now go to the hospital twice a week to be checked," he said. "It was a big problem for me in jail. For three months the doctor didn't know what heart pill I needed. I was very afraid I would die."

"What about your family? Couldn't they help?" I asked.

"My father is dead and my mommy has Alzheimer's. I had to do everything myself. It took time to straighten out my medication. Only my father-in-law tried to help. He asked President Bush to give me a pardon, but the president, he did nothing," Kojima said.

I had so many questions that I didn't know where to start. "I went to the Internet café and read some of the articles on you. There are so many of them," I volunteered.

Kojima looked pleased. "Yes, everybody knows me. The problem is lots of dealers don't like me because I get butterflies cheaper and they're selling very well. You understand? I don't know who, but an American dealer turned me in to Fish and Wild."

"How was your time in prison? Did you have any trouble other than your medication?" I asked.

"It was all right. There were two different Mexican gangs, the Southsiders and the Paisas. The leaders knew why I was in jail. They thought it was funny, crazy, and stupid," he said.

The Chinese mafia also had a presence in California City Prison and took an instant dislike to Kojima.

"They wanted to do something in a fight with me," Kojima revealed. "And I'm not so young boy anymore."

It was the leaders of the Southsiders and the Paisas who protected him. "One of the members was Mexican Japanese, and he tell everybody, 'Oh, that guy has the top half of his pinkie missing.' They automatically thought I was *yakuza*." Kojima laughed as he showed me his pinkie. "Then they scared about me."

After that, the Chinese mafia left him alone and Kojima was treated like royalty. Kojima shrewdly never told them that the *yakuza* cut off the left pinkie. It was his right digit that was missing. He had lost it falling off a ladder as a child.

Kojima next regaled me with stories of homes he owned in Tahiti and Belize. He'd been to them just once, but his son liked to visit more often. That was just the tip of Kojima's real estate empire. He claimed to have purchased his L.A. Mount Olympus house from Joan Collins for one million dollars and sold it for seven million just a few years later. However, his best investment had been a Bel Air home that he'd picked up for a mere seven hundred thousand and then sold for a whopping fifteen million. Finally, his Kyoto residence had an estimated value of one million dollars. When it came to the art of the deal, Kojima liked to pretend he was right up there with Donald Trump.

I knew most of what he said were exaggerations if not downright lies, but that didn't matter. Bits and pieces of the truth were interwoven in his stories as they are in the very best fairy tales.

"I'm so lucky that I sold my Mount Olympus house when I did. If I'd been caught before, Fish and Wild would have taken it from me. Ed Newcomer tried to find all my money in my U.S. bank accounts, but I tricked him. I'd already moved everything back to Japan," Kojima gloated. "That's why I took my Web site down. Fish and Wildlife was always looking on it. I've had to change my password twenty times since I returned home. Yahoo! keeps sending me warnings that someone is trying to break into my e-mail. Ed Newcomer wants to catch me again. He's a really terrible guy."

I commiserated as we walked back to his shop, where we agreed to meet again in two days.

Yuko spoke to me that night about Kojima. "He looks like he could be somebody's grandfather. Do you think if his granddaughter asked that he would stop smuggling butterflies?"

I told her that I doubted it.

Yuko returned to Tokyo with my bountiful thanks, and I continued my assignations with Kojima.

My next visit was on a weekday, and Teramachi Street was crammed with bicycles packed tight as sardines up and down the right-hand side of the street. A small Buddhist temple magically appeared behind a nondescript wall that I hadn't noticed before. It was dwarfed by the bright neon lights of a shop selling comic books, costumes, and lingerie next door.

I rang Kojima's bell, and he quickly came down to let me in. This time, the stench of mildew and sewage enveloped me as soon as I entered his residence.

"I brought a book along. I was hoping you might tell me about some of the butterflies," I said, and placed it on the counter.

Kojima ran his fingers over *Endangered Swallowtail Butterflies of the World*. "That's the butterfly they put me in jail for," he said, gently stroking the Queen Alexandra on the cover.

We opened the book and began to look at the butterflies together. "Did the Japanese authorities penalize you when you finally returned home?"

Kojima gave a contemptuous snort. "They laughed. They say, how come they charge this guy in the airport?" His attention wandered as he pointed to one of the photos. "This butterfly is from Brazil and almost disappeared, but I have a contact and they can get them for me. Brazil is so crazy. No one can collect morphos there anymore, but I have lots."

"How do you get them?" I asked.

"It's easy. My collectors take Brazilian material to another South American country where I have a permit. They mail them from there to me," Kojima revealed. "I do the same thing all over Central and South America. That's why other people cannot get this material but I still can. I have better contacts than they do.

Of course, it made perfect sense. Why would Kojima stop doing what he did best? Smuggling butterflies wasn't a hobby or a vocation; it was part of his DNA.

"I've heard that the same problem exists in India," I said.

Kojima nodded in agreement. "India has nice butterflies. A Tokyo dealer I know used to go there to collect, only he got caught and was jailed for three months. No one can import them now except for me," he divulged with a grin. "I pay and they go out. Two wholesale dealers there send butterflies to me."

He had the same arrangement for procuring rare butterflies and bugs from South Africa, China, Costa Rica, and the rest of the world. Kojima's network was as good as ever. "Today I sold about a thousand dollars of butterflies and bugs on my auction site," he bragged. "I have very good clients."

"You have an amazing business," I agreed.

"That's why Ed Newcomer hate me. He want to kill me. Wait a minute. I can show you some of the butterflies I sell."

He walked into a second room littered with antiques and proceeded up a stairway. All was quiet when I heard a distinct buzz. Looking up, I spied bugs flying about the room. Kojima's bedroom had to be directly above my head. That was where his butterfly-rearing equipment and hatching nets were kept. These were probably parasites that had eaten their way through butterfly pupae upstairs. My skin began to crawl as

Kojima returned with a selection of beetles and butterflies as exotic as their origins.

There were beautiful blue *Morpho rhetenor helena,* iridescent-green *Ornithoptera paradisea,* and dramatic red, black, and cream *Bhutanitis lidderdali.*

"These are from Costa Rica," he said, holding a box toward me.

The bugs inside resembled enameled jewelry, each glistening brightly as if three coats of nail polish had been painted on their backs.

Kojima had brought down something else as well: a photo album filled with publicity shots of his L.A. actor friend. The photos chronicled his career from a young stud to the more mature leading man that he was today.

"I see him on TV all the time," I remarked. "Isn't he married?"

"Yes, but he wants to have kinky sex with me," Kojima confided with a giggle. "I meet so many gay celebrities through him. His wife is very jealous because she hears all the rumors. He's crazy about Oriental boys but doesn't want to touch any of them in California. It's because everybody talking so much. He's a very scared guy. He's always worried about the disease."

Kojima lived on memories as though they were oxygen. I turned the page and saw a much younger Yoshi standing with his arms lovingly draped around the neck of a Caucasian gentleman who appeared to be about fifteen years his senior.

"Is that your father-in-law?" I ventured boldly.

"Yes, but we don't speak anymore. He complain, complain all the time," Kojima said sadly.

"But wasn't he the only one that tried to help when you were arrested?" I pressed.

"He would call me in jail all the time and say that he loved me and worried about my health. Then he suddenly got mad and stopped calling." Kojima's hands fluttered in distress. "My father-in-law is upset, my wife is upset. Only my boy is okay with me."

Kojima was so earnest that I wanted to believe every word he said. "My father-in-law never liked my actor friend in L.A. He knows he's bisexual and was upset because we went on a trip together. He thought we were having an affair."

Probably with good reason. My guess was that it had most likely contributed to their breakup.

I had another hunch to play. "By the way, how were you able to pay for a private lawyer when you were in prison? Did your family cover the expenses?"

"No, no. An antiques dealer in the States lent me the money."

Another piece of the puzzle fell into place as I extracted a sliver of truth from Kojima's lies. Charles Hanson had collected antiques. He'd also lent Kojima money when he was in prison and was angry that he'd never been repaid. It seemed a good bet that Charles Hanson had covered Sasamori's retainer.

I continued to peruse the photo album and found a head shot of Hanson. Kojima snatched it from my hands. "It's a shame about your father-in-law, since you were such good friends. How did you first meet?"

"I worked at my uncle's antiques store when I was in high school and university. I was sixteen years old when my father-in-law first came in. He likes Japanese antiques. He's a very rich man with so many houses and everything. His family is part owner of a big oil company, and they also own a few banks. He's going to leave my boy six hundred million dollars, but I don't want anything." Kojima slipped the photo into the back of the album. "I'm very angry. I don't want to look at him anymore."

"Have you ever been to New York?" I asked, briefly dropping the subject.

"Oh, yes. I had a house there at one time," he replied.

"Really? Where did you live?"

"Trump Tower," he promptly replied. Kojima put the album away. "It stinks in here. We go to the coffee shop?" he offered.

"Where is that smell coming from?" I asked as we walked out the door.

"My brother cooks *kintoki* beans for his business in the back room every night. He uses so much water that the sewer pipes clog up. Water is very expensive here. That's why I go to the communal bath. He must fix it."

I had a feeling that Kojima would be going to the communal baths for a long time to come.

Kojima entertained me with tales of how stupid Customs agents were and how he'd smuggled beetles for years by hiding them on his body and in his luggage. It was almost as if he was giving me lessons. "Carrying insects in hand luggage is okay. Butterflies don't show up on X-ray because they are not meat and bone. That's how I brought two hundred live beetles back from Bolivia. I do it maybe twenty times selling for two million yen. All Customs ever sees is an empty box," he disclosed.

Though I'd already heard them on government undercover tapes, I ate up every story. Kojima spoke of having smuggled film for the *yakuza* in Honolulu, where he bought a seventy-thousand-dollar apartment with the proceeds. He flipped it a few years later for a cool million. There were yearly trips to Paris, Frankfurt, and Florence, and a twenty-five-day luxury vacation with his father-in-law.

"I do so many different things. Once I start something I go crazy. I used to breed fish and birds. I make money but I quit so fast. My life is always up and down, up and down," Kojima explained.

All except for butterflies. They remained the one constant in his life.

"Still, it's too bad that you and your father-in-law aren't in touch anymore," I said, attempting to resuscitate the topic.

Kojima shook his head. "He hates me now. When I come out of jail I have a problem to go home because I have no money. The government took my wallet and credit cards when I'm arrested and give them to my father-in-law. When I get out I need to go home, but they only give me a five-hundred-dollar check. I'm sick and in terrible shape in the heart. I contact my father-in-law, but he refuse to give my credit cards and money back."

"What do you mean? Didn't Immigration and Customs Enforcements agents escort you to the airport and give you a ticket home to Japan?" This was the first I'd heard about it.

"I'm sick when I'm released and they put me in a hospital near LAX for a few days. They said there'd be a ticket waiting for me at the airport, but there wasn't any. How can I go home? The airlines won't take my five-hundred-dollar check!" he cried.

Kojima spoke to a supervisor at Japan Airlines, and they finally agreed to help.

"That's terrible," I said, feeling bad for the man sitting across from me. "But why wouldn't your father-in-law give back your wallet?"

"I don't know. He's upset because my boy had to leave the country. Then after I return home he never call, he never contact me. I don't care about it." Kojima looked at me and smiled. "I'm so happy you're here. I'm glad you found my shop."

We agreed to meet once more before I left Kyoto.

I'd begun to like Kojima. In a strange sense, we'd become friends. Granted, he had yet to discuss his bizarre relationship with Charles Hanson, or admit that he'd probably never been married or that he'd done anything wrong. Most of his stories were lies that he'd told in the past and would tell again. No matter, I was still flattered that he'd taken me into his confidence. Kojima trusted me. How could I betray the man?

We spent my last day in Kyoto together and ended by going to dinner.

"I'm so happy you returned here to see me," Kojima said as we sat down.

"I am, too. You've been a large part of my trip to Japan."

"I try to do the best I can. That is my way," Kojima humbly replied.

"I'm very glad to have stumbled upon your shop," I said, feeling totally guilt-ridden.

Kojima smiled. "Yes, you are lucky. You found a closet queen."

This was his first reference to the topic. "Aha," I pounced. "Are you a closet queen, Yoshi?"

"No, no, no," he protested with a laugh.

It was now a challenge as I tried to catch him off guard. "Was your wife beautiful?" I asked.

He stared at me blankly. "My wife? Which one?"

It was my turn to be surprised. "You only have one wife, right?"

"In the United States?"

Maybe Kojima had as many wives as he supposedly had houses. Something clicked, and Kojima was suddenly himself again. "Who cares about that one?"

"Do you have other wives that I don't know about?" I asked, laughing.

"No, no. Not wife, friend only. Oh my God! Too much trouble. When I call, she always answering and hanging up. Son of a bitch," he nearly spat.

That was another interesting tidbit. Charles Hanson now had a wife, even if in name only. Perhaps Kojima was referring to her.

"Where is your son these days?" I asked.

"He lives in Australia, but he has a girlfriend in Singapore that's pregnant. I also want to see the movie *Australia* because I like Hugh Jackman," Kojima chattered on. "When he's young, he's showing for his naked. There are a lot of nude photos of him on the Internet. He has quite a big one. You can print and put it on your wallpaper."

Good to know, but that wasn't the information I was looking for.

"So, were you ever suspicious of Ed Newcomer?" I asked.

"Such a terrible guy! All the time he's catching someone. It's a terrible problem. One guy told me Ed Newcomer's name and warned me to be careful. But then a young man showed up and said he was Ted Nelson. How can I find out? He wanted to study with me, but I finally kicked him out of my house. That's why he wants to catch me," Kojima ranted.

I silently listened to his side of the story.

"I just speaking on the phone with him about the butterfly Queen Alexandra, and for that they charge me five years," he said. "Then Newcomer tells newspapers that I like him very much. Who cares about that kind of guy? If I do it, I pick up a nice one. Anyway, he cannot touch me anymore now that I'm in Japan."

Kojima felt safe once again. He leaned forward as if to impart a secret. "Maybe somebody kill Ed Newcomer soon."

The bombshell had its intended effect. I sank back in my seat. Could Kojima be serious?

"The Southsider guy say to me, 'If you have a problem and you want, I can do it for you.' He was going to get a message to someone on the outside to kill Newcomer, but I thought about it and say no."

Either Kojima was blowing smoke or he was more dangerous than I had thought.

"That would have been a bad move on your part," I agreed. "Newcomer is a federal agent. They'd track you down, and you'd spend the rest of your life rotting in jail. Are you still in touch with that Southsider guy?"

Kojima shook his head. "No, I don't care about that anymore. I like just selling my butterflies."

"Can anyone from the U.S. or Europe get on your auction site?" I asked, beginning to wonder just who I was dealing with.

"Yes, if they want something, they contact me and I send it Express Mail. I can send under a different name and put a fake address in Japan. That way nobody knows that it comes from me," Kojima answered.

It was a shrewd move on his part. U.S. Customs would never connect the dots.

"You can make maybe four hundred dollars a day. It's easy selling butterflies. You could do too," he said, and then paused. "I still want to do eBay. I have everything I need, but I do not know how to do. How can I use?"

He looked at me expectantly, and a queasy feeling took root in my stomach. Kojima was testing to see if I was game. "I have no idea how to use eBay. I'm pretty bad at that stuff, Yoshi. But you use Internet auction sites in Japan all the time. Shouldn't it be pretty much the same thing?"

Kojima was a clever guy. He'd conducted business over the Internet for years. Putting his material on eBay should be a breeze. That was unless he was looking for his next patsy. He confirmed that with his next request.

"I also want to ask you a favor. Sometimes I need help with a couple American guys that don't want to sell to me American butterflies."

"Why not?" I asked, suspicious of what might be coming.

"Because he's scared for selling outside of U.S. It's nothing do wrong, but some crazy guys don't want to. That's why they sometimes have butterflies that I want but cannot get. Do you have PayPal?"

"No," I replied warily, my mind already conjuring the prison cell I could end up in.

"PayPal is easy. I can send you in one second the money if you buy the butterflies from him. He'll send them to you and then you send them to me by Express Mail. You do nothing wrong because the post office will never open the package," he insisted.

"Is he afraid because you went to prison for selling illegal butterflies?" I asked.

"No, no. He needs permission to send butterflies out of the U.S., but who cares? A lot of other people send to me by Express Mail."

Then why doesn't Yoshi ask one of his other friends in the U.S. to do it? I wondered as a five-alarm fire erupted in my head. *Don't piss him off. Just play along for now. There's still more information to get.* "Sure, no problem."

With that went my fantasy of friendship. Apparently, I was simply another mark. I'd have to end all communication with him. The man was a magnet for trouble.

"Hey, I was wondering if you believe in reincarnation?" I asked him. Kojima was raised Buddhist, and it seemed a fitting question.

"I sometimes have a feeling for it. When I was young I used to feel it a lot but now am getting older," he said.

"Do you think you'll come back a butterfly?" I asked, knowing the answer I hoped to hear.

Kojima didn't disappoint. "I think so."

"That would be fitting," I replied. If karma existed, he'd be hatched by a butterfly breeder and live all of five minutes.

"Only butterflies fade and don't last," he added with a melancholy note.

"I guess they're just like people that way," I reflected. Then another thought struck me. "By the way, what happened to all of your butterflies while you were in prison for those two years?"

Kojima's face turned grave as he stared into space. "I didn't know I'd be gone for so long. I had four thousand butterflies. Some were very expensive and needed to be put in cases with mothballs. When I came home, they'd been eaten by bugs and had all turned to dust. That's why you have to enjoy these things while you're still alive. Because you can't take them to the cemetery." He broke the somber mood with a smile. "Besides, Hugh Jackman naked is better than butterflies, right?"

We walked back to his shop as a light rain began to fall.

"I can't come to the United States anymore, but maybe if you go on a trip to South America, I can go with you," he offered. "Anyway, I'm going to see you the next time you come."

He unlocked the shop door. There was no sign of his brother, no scent of *kintoki* beans cooking in the back room, and the second bicycle remained untouched where it sat parked. I'd never know exactly what was true and what was a lie. The only reality was that Kojima remained the king of butterflies.

He handed me a plastic umbrella for my walk home.

"You won't forget me?" I asked as I stood at the door.

He gave me a kiss on both cheeks. "I won't forget you," he promised.

I wouldn't forget him, either. I think of Kojima every time I see a butterfly and say a silent prayer of thanks that it's still flying free.

Acknowledgments

Thanks go to numerous people without whom this book would never have undergone the metamorphosis from idea to the written word. My former editor, Sarah Durand, asked what I wanted to write after my final book in the Rachel Porter mystery series. She received her answer over martinis at a conference in Alaska. Sarah championed my proposal and helped to make sure it was green-lit. Then she left for another publishing house. Kate Hamill, my second editor at William Morrow, cheerfully adopted my manuscript. She wisely guided me by emphasizing the heart of the story. Kate moved on to a different imprint one week before the manuscript was due. Tessa Woodward gamely came on board as the third editor of the project. She deftly pointed out what still needed to be added and helped with the final shaping of the book. I owe each of these women a debt of gratitude.

Ed Newcomer is the heart and soul of this story and I am deeply grateful that he willingly gave so much of his time. Time is one commodity that U.S. Fish and Wildlife Service (USFWS) agents don't have enough of. Ed patiently put up with my prodding as I proceeded to dig into every corner of his life. He has requested that if the book ever becomes a film, the actor playing his role should be chiseled with rock-hard abs. I promise to do whatever I can if that happens. A note of thanks also goes to his wife, Allison, for allowing me into their lives and home.

Retired USFWS agent Sam Jojola first brought this case to my attention. He's been a good friend and supporter for many years, and I can't thank him enough for all of his help on this book and so many other projects. He has dedicated his life to fighting the illegal wildlife trade and continues to do so.

The Law Enforcement Division of the USFWS has some of the most dedicated agents, inspectors, and employees on earth. My respect for these men and women knows no bounds. I want to thank Erin Dean, Marie Palladini, John Brooks, Lisa Nichols, Mike Osborne, Chris Nagano, Sandra Cleva, Valerie Fellows, and Tamara Ward for all of their help. In addition, Assistant U.S. Attorney Joseph Johns provided invaluable information on the case.

No book about butterflies would be complete without the assistance of lepidopterists. I am especially grateful to Dr. Gordon Pratt whose love for, and knowledge of, butterflies is boundless. I also want to thank Dr. John Emmel for his perceptive insight on butterflies and butterfly collectors, along with UC-Riverside entomologists Ken Osborne, Greg Ballmer, and Jeremiah George. Brent Karner, of the Natural History Museum of Los Angeles County, always answered my calls for help.

I knew no one when I flew to Japan, but I left with two good friends who were always there for me. Many thanks go to Walter Krumholz. He made me feel at home and also introduced me to the indispensable Yuko Tsukada. Yuko was my guide not only to Kyoto but also to some of the mysteries of Japan. She taught me not to be such a *gaijin*.

Numerous friends contributed in countless ways to this book. I especially want to thank Regina Banks, Julie Holzworth, Janet Haller, Bridget Bly, and Dinah and Stephen Lefkowitz for their help and support. Anna Collins read my manuscript with a critical eye and offered pointed suggestions while listening to me wail. She also provided many necessary glasses of wine. Friend and author Lori Andrews was the first to convince me that this story had the merit to be a good book. Many thanks to George Brenner for scrupulously reading each chapter and doing his best to keep me sane.

Finally, I'd like to thank my agent, Dominick Abel. He trusted that a mystery author could follow her passion and write a nonfiction book.

Q&A with Jessica Speart about *Winged Obsession*

1. How did you first hear about Yoshi Kojima and his elusive butterfly smuggling?

The story became national news soon after Kojima's arrest. It was just such a quirky topic. Who would smuggle butterflies and why? Most people envision butterfly collectors as mild-mannered lepidopterists wearing pith helmets and swinging their nets wildly in the air. *Saturday Night Live* even used Kojima's arrest as fodder for a skit. What people don't realize is that the illegal butterfly trade translates into big money. There are butterflies that sell for $39,000 and up. U.S. Fish and Wildlife estimates the illegal butterfly trade to be worth $200 million a year.

2. What did you expect to happen when you flew across the world on a whim to find him?

I honestly had no idea what would take place. Fish and Wildlife agent Ed Newcomer was no longer in touch with Kojima, and I had no contacts in Japan to check his whereabouts. For all I knew, he would be out of town when I arrived. Kojima constantly traveled before his arrest. One week he'd be collecting butterflies in Madagascar and the next would find him in Costa Rica. I figured there was a fifty-fifty chance he'd be at home. Even so, I wouldn't have been surprised to knock on his door only to have him slam it closed in my face. As it turns out, Kojima had been hospitalized for three weeks and released just four days before my arrival.

3. Do you ever fear the repercussions, if Kojima were to learn your true identity?

I fully expect that Yoshi will figure out who I am. He's a very clever man. I interviewed a number of his friends and acquaintances in California. Word will eventually get back to him, if it hasn't already. As far as repercussions, that's not something I tend to dwell on. I'm sure Kojima will be angry at first, but he also craves attention. "You're the butterfly movie star," Newcomer once told him. Kojima had replied that he hoped so. He still sells butterflies on the Internet. Who knows? This book could make him more popular than ever. In that sense, it's probably a double-edged sword.

4. The relationship between Kojima and Newcomer develops into something completely bizarre. Do you think this played a large role in Kojima's eventual capture and imprisonment?

Absolutely. Kojima flew to L.A. in July 2006 solely for one reason. He'd become as obsessed with Ed Newcomer as he was with butterflies, and Newcomer gladly provided the bait. In the end, Kojima's heart betrayed him as much as Newcomer did.

5. It seems like Kojima spent his whole life weaving a web of lies. What is it about this man that makes him able to get away with such deceit?

It's amazing. He was able to fool everyone around him. It didn't matter if they were government officials or some of his closest friends. I think Yoshi had lied for so long that he'd actually begun to believe his own tales. One butterfly collector remarked that Yoshi reminded him of Frank Abagnale, the con man in *Catch Me If You Can*. Kojima was that good until he got caught. I heard many of the same stories when we met in Japan. The only difference was that after prison, Kojima had even more tales to weave.

6. Tell us about the theme of "obsession" and the role it plays throughout the story you uncovered.

I began my research for this book by interviewing lepidopterists and butterfly collectors. The consistent thread throughout was how their interest in collecting had morphed into full-blown obsession, creating havoc in their lives. Most collectors were men who admitted to having problems with social skills. Their obsession had led many to divorce, bankruptcy, depression, alcoholism, and drugs. One collector confessed, "None of us are normal. We're driven toward something that doesn't put food on the table and that we can't take with us when we die. There's a lot about this hobby that goes way too far. There is a dark side here."

Vladimir Nabokov, author of *Lolita*, referred to his obsession with butterflies as his demon. In the same vein, Kojima had to have every single species of butterfly and as many of them as he could get. The more rare it was, the more he wanted it.

Finally, there was my own obsession with the story. The more I learned about Kojima, the more I wanted to know. That obsession took me halfway around the world in order to find him.

7. Were there any incidents with Kojima that weren't covered in *Winged Obsession*?

What readers don't know is that I maintained a Skype relationship with Kojima upon returning home. Yoshi was very interested in purchasing a particular species of butterfly native to the United States. The *Diana fritillary* is a large, beautiful butterfly found in moist, mountainous habitats in the southern Appalachians and the Ozark Mountains. The female is a dramatic dark blue-black, with a dusting of light blue spots on the edge of its wings, and is sought after by collectors. The population is declining over much of its historic range and has been listed as a species of concern.

Yoshi had been in touch with an American collector who didn't have the proper U.S. Fish and Wildlife permits to send the *Diana fritillary* overseas. The collector was probably also wary of dealing with Kojima. That's where I came in. Yoshi asked if I would act as an intermediary. He wanted me to buy the butterflies from the collector and then send them to him in Japan by Express Mail. Yoshi kept assuring me that I wouldn't get in trouble since authorities rarely inspect Express Mail packages. The more I hedged, the more Kojima pressed that I do him this favor. It was then that I ended our Skype conversations.